Chemistry 102
Lecture Notes

Dana S. Chatellier

University of Delaware

KENDALL/HUNT PUBLISHING COMPANY
4050 Westmark Drive   Dubuque, Iowa 52002

DEDICATION

My first book, <u>Chemistry 101 Lecture Notes</u>, was dedicated to a lot of people, but didn't name any of them. So, this time, I decided to name names. (Or at least <u>one</u> name...)

I first compiled my lecture notes while teaching the equivalent of Chemistry 102, Chemistry 136, at Willamette University in Salem, Oregon in the spring of 1985. One of my students in that first class was a young man by the name of Brad Aitchison. After about a week or two, I knew I could count on Brad to ask at least two or three good, insightful, penetrating questions about whatever the subject matter for the day happened to be. Brad was never unprepared for class in any way, and he taught me, right then and there, that <u>I</u> could never hope to get away with being unprepared for class in any way. His inquisitiveness, enthusiasm, and relentless pursuit of the truth literally made me into a better teacher.

Since then, I have had the opportunity to work with many other students. Some have had minds as keen and devoted as Brad's, and some have not. What I have learned about myself, in the course of working with both kinds of students, is that I am only as good a teacher as my students are motivated to learn. Most teachers enjoy working with bright and talented students, but I have found that I am also inspired by the students who do not possess great aptitude for chemistry, but who doggedly strive to do the best that they can possibly do, sometimes under severely handicapping conditions. I still have difficulty working with students who are both unmotivated and unskilled, which only shows that I still have a great deal to learn about teaching.

This book is lovingly dedicated to Brad, and to all the students like him with whom it has been my privilege to work over the last few years. They made me what I am today, and I thank them.

--Dana S. Chatellier
September, 1989

ISBN 0-7575-1509-6   ISBN 978-0-7575-1509-5

TABLE OF CONTENTS

PREFACE

This book is a typewritten version of the lecture notes which I have used for the last several years in teaching Chemistry 102 at the University of Delaware. Chemistry 102 is the second-semester course of a two-semester sequence in chemistry intended largely for non-science majors. Most general chemistry textbooks fall into one of two categories: those intended for a one-semester course for non-science majors, or those intended for a two-semester sequence for science majors. As the demands of Chemistry 101 and Chemistry 102 fall somewhere between these two extremes, it was deemed useful -- indeed, necessary -- for a good set of lecture notes to accompany the textbook for the course, the better to assist the student in approaching the study of chemistry for what may be the first time.

My original lecture notes were compiled during my first teaching position, at Willamette University in Salem, Oregon. They were adapted from the text in use at that time at that institution -- "Fundamentals of Chemistry" by James E. Brady and John R. Holum. Since then, they have undergone considerable revision and alteration. In preparing this manuscript, I have attempted to make the material therein adaptable to many of the existing general chemistry textbooks, including the one currently in use in Chemistry 102 at the University of Delaware -- namely, "General Chemistry" by Henry F. Holtzclaw, Jr. and William R. Robinson. The order in which topics are presented is similar to that in most current general chemistry textbooks. However, the modular format employed -- that is, one topic is covered per page in most instances -- allows for facile rearrangement of topics simply by re-ordering pages.

This book is not a textbook. Indeed, it would be difficult to use it as such -- there are very few illustrations, and not very many problems for students to practice working. This was done intentionally, since this book is intended as a supplement to a general chemistry textbook, most of which have illustrations and practice problems galore. The advantage to the student of having this book is that it is written in a style which is more conversational than that found in most textbooks -- in short, it's written the way I teach it! Many of my students have found this feature to be very beneficial. In addition, this book allows the student to simply listen to my lectures without the bother of taking notes, if such is the student's desire. For some students, this may be the best way to learn!

This book will obviously find its greatest applicability in courses in general chemistry which I teach, but it is hoped that students in other courses (and possibly at other institutions) will find it useful as well. Chemistry is a fascinating field of study, and one whose importance in the modern technological world cannot be denied. If this book helps people understand the world and the age in which they live, then it will have served its purpose.

--Dana S. Chatellier
September, 1989

TO THE STUDENT:

A lot of students have a great deal of difficulty with Chemistry 102. Part of this is probably due to the fact that the material _is_ a bit more difficult than the material in Chemistry 101 (as is expected for a higher-numbered course!). However, another part is probably due to what Dr. Frank Westheimer of Harvard University calls "the vertical nature of learning in science".[1] According to Dr. Westheimer, "education in science is highly vertical, where one subject is built upon knowledge of another...Biochemistry cannot be understood by anyone who doesn't have a minimal knowledge of organic chemistry, and organic chemistry is hard to explain to someone who doesn't know some general chemistry."[1] To expand upon Dr. Westheimer's words, Chemistry 102 is hard to explain to someone who doesn't _know_ (note the heavy emphasis on the word _know_) Chemistry 101. Some students find themselves re-learning Chemistry 101 in Chemistry 102, which only compounds the difficulty of an already difficult curriculum.

If you find yourself having difficulty with Chemistry 102, see one of your instructors. You'll find that they really _do_ want to see you succeed. Take advantage of all of the resources available to you -- tutors, help sessions, flash cards, worked-out examples in your textbook -- and, oh, yes, these lecture notes. Feel free to make notes to yourself in the margins and between lines of this book as needed. These lecture notes were written with _you_ in mind -- I hope they help.

Best wishes for a rewarding semester learning about chemistry.

--Dana S. Chatellier
September, 1989

---

1) From "Education of the Next Generation of Nonscientists" by Dr. Frank H. Westheimer. _Chemical & Engineering News_, July 4, 1988 issue, pp. 32-38.

MISCIBILITY OF LIQUIDS:

As the saying goes, "oil and water don't mix." This is fortunate for the makers of salad dressing! The makers of alcoholic beverages, on the other hand, take advantage of the fact that ethyl alcohol and water do mix together. The tendency of some liquids to "mix" -- that is, to form homogeneous mixtures -- is referred to as their miscibility. Two liquids are miscible if they are each soluble in the other, regardless of the relative proportions used.

Why are some liquids miscible and other liquids are not miscible? The answer has to do with the intermolecular forces of attraction between molecules of the liquids in question. For two liquids to be miscible, their intermolecular attractive forces must be of approximately equal strength. Any attempt to mix two liquids whose intermolecular attractive forces are of unequal strengths will result in the molecules of the liquid with the stronger intermolecular attractive forces tending to form relatively strong bonds to each other, leaving the molecules of the liquid with the weaker intermolecular attractive forces to form a separate layer. In short, no mixing takes place.

A simple rule which summarizes all of this is the "like dissolves like" rule, which says that liquids of similar polarity are miscible with each other, while liquids with different polarities are not miscible. For example, water (H_2O) is a polar molecule -- water molecules tend to connect to each other by forming hydrogen bonds. Liquids such as ethyl alcohol (CH_3CH_2-O-H), which also tend to form hydrogen bonds, are generally miscible with water. However, liquids which do not tend to form hydrogen bonds (such as the "oil" in salad dressing) are usually immiscible with water. (The "oil" in salad dressing is composed largely of carbon and hydrogen, which are similar in electronegativity. Thus, "oil" is nonpolar, and its molecules are attracted to each other by relatively weak London forces.)

SOLUBILITY OF GASES IN LIQUIDS:

Gases are soluble in liquids to some extent, as a moment's thought will make obvious. For example, oxygen (O_2, a gas) is transported to the brain by the blood (a liquid). The solubility of a gas in a liquid depends mostly on two properties of the gas in question -- its _pressure_ and its _temperature_.

Not surprisingly, the solubility of a gas in a liquid _increases_ as the _pressure_ of the gas _increases_. Again, this should be obvious from everyday experience. For example, when you open a can of a "carbonated" beverage, you see bubbles come to the surface of the liquid. The bubbles are made of carbon dioxide (CO_2, a gas) which was dissolved in the liquid until you opened the can. Cans of carbonated beverages are sealed so as to contain a slightly higher pressure of gas than atmospheric pressure. Exposing the carbonated beverage to a lower pressure (atmospheric pressure) causes the dissolved gas (CO_2) to be less soluble in the liquid than it was, with the result that bubbles form.

Somewhat surprisingly, the solubility of a gas in a liquid _decreases_ as the _temperature_ of the liquid _increases_. (For _solids_, the opposite is true -- for example, sugar (a solid) is more soluble in hot water than in cold water.) Once again, you've probably seen this phenomenon before. For example, if you heat a pan of water on a stove, you will notice that just before the water boils, some relatively small bubbles will form along the insides of the walls of the pan. These bubbles are not made of water vapor (like the "boiling bubbles" are), but are actually made of oxygen gas which had been dissolved in the water before you heated it. This phenomenon is the basis for the environmental problem known as _thermal pollution_ -- if the temperatures of the lakes and ponds where fish live rise above certain levels, the amount of oxygen dissolved in the water will be reduced below the point at which the water can support fish and other marine life.

SOLUBILITY OF SOLIDS IN WATER:

When a crystalline solid dissolves in water, it does so because water molecules strike the crystal lattice with enough force to dislodge particles of the solid, whether these particles are <u>molecules</u> (such as the $C_{12}H_{22}O_{11}$ molecules of which sugar is made) or <u>ions</u> (such as the Na^+ and Cl^- ions of which table salt -- sodium chloride -- is made). The water molecules then surround the particles of the solute, forming a "solvent cage" around them which is held in place by ion-dipole or dipole-dipole attractive forces. These "solvent cages" keep the solute particles separated from each other, preventing recrystallization of the solid.

Not all ionic compounds are soluble in water. The reasons for this are not well understood, but a series of empirical rules for predicting whether an ionic compound is soluble in water has been developed. The <u>solubility rules</u> given below are numbered 1 through 6 for future reference.

The following ionic compounds are <u>soluble in water</u>:

1. All salts containing Na^+, K^+, or NH_4^+ ions.

2. All salts containing NO_3^-, ClO_4^-, ClO_3^-, or $C_2H_3O_2^-$ ions.

3. All salts containing halide ions (Cl^-, Br^-, I^-) <u>except</u>:
    Ag^+, Hg_2^{2+}, and Pb^{2+} halides.

4. All salts containing sulfate (SO_4^{2-}) ions, <u>except</u>:
    $CaSO_4$, $BaSO_4$, $RaSO_4$, $SrSO_4$, $PbSO_4$, and Hg_2SO_4.

The following ionic compounds are <u>insoluble in water</u>, but are salts of weak acids and are therefore <u>soluble in acid</u>:

5. All salts containing oxide (O^{2-}) and hydroxide (OH^-) ions, <u>except</u>:
    Ca^{2+}, Ba^{2+}, Na^+, K^+, and NH_4^+ oxides and hydroxides.

6. All salts containing CO_3^{2-}, PO_4^{3-}, S^{2-}, or SO_3^{2-} ions, <u>except</u>
    those which are soluble according to Rule # 1 above.

7. Insoluble Compounds under rule #5 and #6

3

USING THE SOLUBILITY RULES:

Problem: Convert pure, crystalline $CaCO_3$ into pure, crystalline $CaCl_2$.

Solution: $CaCO_3$ is insoluble in water, but soluble in acid (Rule # 6). Therefore, dissolve the $CaCO_3$ in acid to get Ca^{2+} ions into solution. A good acid to choose would be HCl, since this will also get Cl^- ions into solution. $CaCl_2$ is soluble in water (Rule # 3), so simply heating this solution until all of the water has boiled away will leave a residue of pure, crystalline $CaCl_2$. The equation of the reaction which occurs is: $CaCO_3 + 2\ HCl \longrightarrow CaCl_2 + H_2O + CO_2$. (Carbonates (CO_3^{2-}) give off bubbles of CO_2 gas whenever they react with acids.)

Problem: Separate a mixture of $MgCl_2$ and $BaCl_2$, obtaining pure $MgCl_2$ and pure $BaCl_2$ at the end of the separation process.

Solution: Both $MgCl_2$ and $BaCl_2$ are soluble in water (Rule # 3). Thus, dissolving them in water produces a solution which contains Mg^{2+} ions and Ba^{2+} ions. These ions can be separated by adding NaOH to the solution. $Mg(OH)_2$ is insoluble in water (Rule # 5), so it emerges from the solution as a solid. The Ba^{2+} ions remain in solution, since $Ba(OH)_2$ is soluble in water (Rule # 5). The $Mg(OH)_2$ can be isolated, acidified with HCl, and heated to dryness to give pure $MgCl_2$. The remaining solution can be acidified with HCl and boiled to give pure $BaCl_2$.

Problem: An unknown crystalline solid may contain any or all of the following: $Ca(NO_3)_2$, $CuCl_2$, K_2CO_3, SrI_2. Adding water to the unknown solid gave a solid residue. Treating this residue with H_2SO_4 caused the residue to dissolve completely and produced bubbles of a gas. Identify the unknown solid.

Solution: The bubbles suggest that the residue is a water-insoluble carbonate salt, but it can't be $CaCO_3$ or $SrCO_3$, since either of these would react with H_2SO_4 to produce water-insoluble sulfate salts (Rule # 4). Therefore, it must be $CuCO_3$, and the unknown solid was a mixture of $CuCl_2$ and K_2CO_3.

ELECTROLYTES:

As you probably already know, it's not a good idea to go swimming in the ocean during a thunderstorm! Neither is it a good idea to drop an electrical appliance into your bathtub while you're bathing! This is because both the ocean and your bath water conduct electricity, which could harm you if you're careless.

Pure water does <u>not</u> conduct electricity, but water which contains <u>ions</u> <u>does</u> conduct electricity. Any substance which produces ions when dissolved in water, thereby creating a solution which conducts electricity, is called an <u>electrolyte</u>. All water-soluble <u>ionic</u> compounds (such as NaCl in salty ocean water) are electrolytes, but some water-soluble <u>covalent</u> compounds are also electrolytes. An example is <u>HCl</u>, which is normally a <u>gas</u>, but reacts with water to form ions:

$$HCl_{(g)} + H_2O_{(l)} \longrightarrow H_3O^+_{(aq)} + Cl^-_{(aq)}$$

Some covalent compounds are <u>nonelectrolytes</u> -- their aqueous solutions do not conduct electricity. A common example of a nonelectrolyte is table sugar (sucrose, $C_{12}H_{22}O_{11}$), whose aqueous solutions contain sucrose <u>molecules</u>, but not many ions.

Compounds such as NaCl and HCl are examples of <u>strong</u> electrolytes -- their aqueous solutions are good conductors of electricity. Some covalent compounds which are electrolytes produce aqueous solutions which are relatively <u>poor</u> conductors of electricity. These compounds are called <u>weak</u> electrolytes. Examples include ammonia (NH_3) and acetic acid ($HC_2H_3O_2$, found in vinegar). The reactions these compounds undergo when dissolved in water are shown below.

$$NH_3{}_{(g)} + H_2O_{(l)} \longrightarrow NH_4^+{}_{(aq)} + OH^-_{(aq)}$$
$$HC_2H_3O_2{}_{(aq)} + H_2O_{(l)} \longrightarrow H_3O^+{}_{(aq)} + C_2H_3O_2^-{}_{(aq)}$$

Compounds such as ammonia and acetic acid are <u>weak</u> electrolytes because the reactions shown above do <u>not</u> go to completion -- a relatively <u>small</u> number of ions is formed, resulting in a solution which is a relatively <u>poor</u> conductor.

COLLIGATIVE PROPERTIES:

In the winter, when snow and ice make the highways slippery and thus dangerous for travel purposes, rock salt is spread out over the roads in order to "melt" the ice. Actually, the salt doesn't "melt" the ice -- that is, it doesn't supply any additional heat. What happens is that the rock salt dissolves in the ice, forming a solution whose freezing point (or melting point, if you like) is lower than the freezing point of pure water. Since colder temperatures are needed for freezing to occur, the ice-salt mixture melts more readily than pure ice does.

The lowering of the freezing point of an aqueous solution of a salt is one example of several properties of solutions referred to collectively as colligative properties. What distinguishes a colligative property from any of the other properties of solutions is that colligative properties depend only on the numbers of particles of solute and solvent present in the solution, and not on the identities of the solute and the solvent. For example, the freezing point of any aqueous solution will be lower than the freezing point of pure water -- it doesn't matter whether the solute is NaCl, $CaCl_2$, $C_{12}H_{22}O_{11}$, or any other solid. However, the identity of the solute does help determine the extent to which the freezing point of the solution is lowered.

The reason for the lowering of the freezing point of a solution is that the presence of the solute particles makes it more difficult for the particles of the solvent to arrange themselves into the crystalline lattice formation needed for freezing to occur. More energy must be removed from these solvent particles -- that is, lower temperatures must be achieved -- before crystallization (freezing) can take place. Similarly, the boiling point of a solution is higher than the boiling point of the pure solvent, since the solute particles occupy some of the surface area of the solution, making vaporization of solvent less likely to occur.

6

MOLALITY AND FREEZING POINT DEPRESSION:

The extent to which a solute which is a <u>non-electrolyte</u> lowers the freezing point of a solution is given by the equation: $\Delta T = k_f m$, where ΔT is the freezing point depression, k_f is the <u>freezing point depression constant</u>, and m is the concentration of the solution, expressed as its <u>molality</u>. The molality of a solution is defined as the <u>amount of solute (in moles) per kilogram of solvent</u> present in the solution. (Spelling is very important here -- do not confuse mola<u>r</u>ity (moles/liter, symbol = M) with mola<u>l</u>ity (moles/kilogram, symbol = m).)

<u>Problem</u>: What is the molality of a solution prepared by dissolving 34.23 grams of sucrose (table sugar, $C_{12}H_{22}O_{11}$) in 50.00 mL of water?

<u>Solution</u>: First, find the molecular weight of sucrose:

M.W. sucrose = $(12 \times 12.011) + (22 \times 1.0079) + (11 \times 15.9994)$ = <u>342.30 grams/mole</u>.

Next, find the amount of sucrose which dissolved (in moles):

Amount of sucrose = $\frac{34.23 \text{ grams sucrose}}{342.30 \text{ grams/mole}}$ = <u>0.1000 moles sucrose</u>. (4 sig. figs.!)

Next, find the <u>mass</u> of water used, in <u>kilograms</u>: (density = 1.00 g/mL)

Mass of water = 50.00 mL $\times$ (1.000 gram/mL) $\times$ (0.001000 kg/gram) = <u>0.05000 kg water</u>.

Finally, compute the molality by dividing the amount of the solute used (in moles) by the mass of the solvent used (in kilograms):

Molality of Solution = $\frac{0.1000 \text{ moles sucrose}}{0.05000 \text{ kg water}}$ = <u>2.000 molal</u> = <u>2.000 m</u>.

<u>Problem</u>: What would be the freezing point of the above solution? The value of k_f for water is 1.86 ^oC/molal.

<u>Solution</u>: Use the above equation to find the freezing point depression:

Freezing Point Depression = $\Delta T = k_f m$ = (1.86 ^oC/molal) $\times$ (2.000 molal) = <u>3.72 ^oC</u>.

Now, since this represents a freezing point <u>depression</u>, <u>subtract</u> this number from the freezing point of pure water to get the solution's freezing point:

Freezing Point of Solution = 0.00 ^oC - 3.72 ^oC = <u>-3.72 ^oC</u>.

BOILING POINT ELEVATION:

The extent to which a solute which is a non-electrolyte raises the boiling point of a solution is given by an equation which is very similar to the equation for determining the freezing point depression of a solution. The equation is: $\Delta T = k_b m$, where ΔT is the boiling point elevation, k_b is the boiling point elevation constant, and m is the molality of the solution.

Problem: Maple syrup has a boiling point of 104.6 $^{\circ}C$. Assuming that maple syrup is essentially a solution of sucrose ($C_{12}H_{22}O_{11}$, M.W. = 342.30 g/mole) in water, calculate the molality of maple syrup. k_f for water = 0.51 $^{\circ}C$/molal.[1]

Solution: Pure water has a boiling point of 100.0 $^{\circ}C$, so the boiling point elevation can be found by subtracting this from maple syrup's boiling point:

Boiling Point Elevation = ΔT = 104.6 $^{\circ}C$ - 100.0 $^{\circ}C$ = 4.6 $^{\circ}C$.

Now, rewriting the above equation as: $m = \Delta T / k_b$, we can easily find the molality:

Molality = $\Delta T / k_b$ = (4.6 $^{\circ}C$) ÷ (0.51 $^{\circ}C$/molal) = 9.0 molal.

Problem: What is the percentage by mass of sucrose in maple syrup?[1]

Solution: From the above problem, maple syrup is 9.0 molal sugar in water -- that is, 9.0 moles of sugar are present for every kilogram of water that is present. Converting 9.0 moles of sugar (sucrose) into kilograms, we get:

9.0 moles sucrose x $\frac{342.30 \text{ grams}}{1 \text{ mole}}$ x $\frac{1 \text{ kilogram}}{1000 \text{ grams}}$ = 3.1 kg sucrose.

The percentage by mass is simply the mass of sucrose divided by the total mass of the solution, expressed as a percentage:

Total Mass = 3.1 kg sucrose + 1.0 kg water = 4.1 kg solution.

Percentage by Mass = $\frac{3.1 \text{ kg sucrose}}{4.1 \text{ kg solution}}$ x 100% = 76% sucrose by mass.

(The next time you eat pancakes and pour a lot of maple syrup on them, you might stop to consider the fact that maple syrup is about three-fourths pure sugar!)

---

1) From a similar problem on page 255 in "Chemical Principles and Properties" by M. J. Sienko and R. A. Plane. 1974 by McGraw-Hill, Inc.

VAPOR PRESSURE LOWERING AND MOLE FRACTION:

The boiling point of a liquid is the temperature at which the liquid's vapor pressure equals the pressure of the surrounding atmosphere. An aqueous solution of a non-volatile solid has a higher boiling point than the boiling point of pure water. This is due to the fact that the solution's vapor pressure is <u>lower</u> than the vapor pressure of pure water. Therefore, more heat is needed to raise the solution's vapor pressure to atmospheric pressure than is needed to raise the vapor pressure of water to atmospheric pressure. Aqueous solutions of non-volatile solids have lower vapor pressures than water because some of the surface areas of these solutions are occupied by particles of the dissolved solutes, making less of the surface area available to be occupied by water molecules. Since vapor pressure is caused by the evaporation of water molecules at the <u>surface</u> of a liquid, the fewer the water molecules at the surface, the lower the liquid's vapor pressure.

The extent to which a non-volatile solute lowers the vapor pressure of a solution depends upon the <u>mole fractions</u> of the solute and solvent present. The <u>mole fraction</u> of one component in a solution is simply the amount of that component present (in moles) divided by the <u>total</u> amounts of all of the components present in the solution (in moles). Mole fractions can be calculated as shown below.

<u>Problem</u>: Alcoholic beverages which are "150 proof" contain 75% ethyl alcohol and 25% water by volume. Calculate the mole fraction of each component in this solution. (Ethyl alcohol: density = 0.79 g/mL, M.W. = 46.07 g/mole.)

<u>Solution</u>: 100.0 mL solution = 75.0 mL alcohol + 25.0 mL water. Moles: Amount of alcohol = (75.0 mL) x (0.79 g/mL) ÷ (46.07 g/mole) = <u>1.3 moles alcohol</u>. Amount of water = (25.0 mL) x (1.00 g/mL) ÷ (18.02 g/mole) = <u>1.4 moles water</u>.

Total moles = 1.3 moles alcohol + 1.4 moles water = <u>2.7 total moles</u>.

$$X_{alcohol} = \frac{1.3 \text{ moles alcohol}}{2.7 \text{ total moles}} = \underline{0.48}. \qquad X_{water} = \frac{1.4 \text{ moles water}}{2.7 \text{ total moles}} = \underline{0.52}.$$

RAOULT'S LAW:

The extent to which a non-volatile solute lowers the vapor pressure of a solution is given by Raoult's Law, which can be written in equation form as: $P_{solution} = P_{solvent}X_{solvent}$, where $P_{solution}$ is the vapor pressure of the solution, $P_{solvent}$ is the vapor pressure of the pure solvent, and $X_{solvent}$ is the mole fraction of the solvent present in the solution. A solution which obeys Raoult's Law is called an ideal solution, just as an ideal gas obeys PV=nRT.

Problem: The vapor pressure of pure water is 30.0 Torr at 29 $^{\circ}$C. What is the vapor pressure of a solution made by dissolving 90.08 grams of glucose ($C_6H_{12}O_6$, M.W. = 180.16 g/mole) in 90.08 mL of water at this temperature?

Solution: Start by converting the measured quantities into amounts:
Amount of water = (90.08 mL) x (1.00 g/mL) ÷ (18.02 g/mole) = 5.00 moles of water.
Amount of glucose = (90.08 g) ÷ (180.16 g/mole) = 0.5000 moles of glucose.

Now, find the mole fraction of water present in the solution:
Total Moles = 5.00 moles of water + 0.5000 moles of glucose = 5.50 total moles.
$$X_{water} = \frac{moles\ water}{total\ moles} = \frac{5.00\ moles\ water}{5.50\ total\ moles} = 0.909.$$

Finally, simply use Raoult's Law to find the solution's vapor pressure:
$$P_{solution} = P_{solvent}X_{solvent} = P_{water}X_{water} = (30.0\ Torr) \times (0.909) = 27.3\ Torr.$$

Problem: The same as above, only using 29.22 grams of sodium chloride (NaCl, M.W. = 58.44 g/mole) as the solute. What is the solution's vapor pressure?

Solution: At first glance, the calculations appear to be the same, since 0.5000 moles of NaCl is used. However, NaCl is ionic -- one mole of NaCl produces one mole of Na^+ ions and one mole of Cl^- ions when dissolved. Hence:
Total Moles = 5.00 moles water + 0.5000 moles Na^+ + 0.5000 moles Cl^- = 6.00 total moles.
$$X_{water} = \frac{moles\ water}{total\ moles} = \frac{5.00\ moles\ water}{6.00\ total\ moles} = 0.833.$$
$$P_{solution} = P_{solvent}X_{solvent} = P_{water}X_{water} = (30.0\ Torr) \times (0.833) = 25.0\ Torr.$$

10

OSMOTIC PRESSURE:

All living things are made of cells, which contain water in addition to various other substances. Water molecules are the smallest molecules present in living cells -- in fact, they are so small that they are able to pass through the "pores" in the membrane which surrounds the cell. Cell membranes are sometimes called semipermeable membranes for this reason -- they allow some molecules (such as water molecules) to pass through, but not others. The flow of water molecules (or molecules of any other solvent) through a semipermeable membrane is called osmosis. Generally, osmosis tends to occur in such a way as to equalize the concentrations of the solutions on opposite sides of the membrane in question -- that is, water tends to flow from a less concentrated solution into a more concentrated solution if the two solutions are separated by a porous membrane. This process can be stopped or reversed by increasing the pressure inside the more concentrated solution. (One way to obtain pure water from salt water makes use of this principle. This process is called reverse osmosis.) The pressure required to exactly stop osmosis from a pure solvent into a solution is called the osmotic pressure of that solution. The osmotic pressure of a solution can be calculated using the equation: $\Pi = MRT$, where Π (capital Greek letter pi) is the osmotic pressure, M is the molar concentration of the solution, R is the ideal gas constant $(0.082056 \frac{L \ atm}{K \ mole})$, and T is the Kelvin temperature.

Problem: What is the osmotic pressure of a 1.00 M solution at 25 $^{\circ}$C ?

Solution: T = 25 + 273 = 298 K. Now, just use the above equation:

$$\Pi = MRT = (1.00 \frac{mole}{L}) \times (0.082056 \frac{L \ atm}{K \ mole}) \times (298 \ K) = \underline{24.5 \ atm}.$$

(Note: It can be shown that the osmotic pressure above is sufficient to support a column of water over 800 feet tall! This helps to explain how water is able to flow to the leaves at the tops of very tall trees.)

USING COLLIGATIVE PROPERTIES TO DETERMINE MOLECULAR WEIGHTS:

Problem: A solution was prepared by dissolving 5.00 g of an unknown protein in 50.00 mL of water. The resulting solution had an osmotic pressure of 1.00 Torr at a temperature of 27 $^{\circ}$C. What is the molecular weight of the protein?

Solution: Using $\Pi = MRT$, find the concentration of the solution:

$M = \Pi/RT = (1.00 \text{ Torr}) \times (1 \text{ atm}/760 \text{ Torr}) \div [(0.082056 \frac{L \text{ atm}}{K \text{ mole}}) \times (27 + 273 \text{ K})]$

$= 5.35 \times 10^{-5}$ moles/liter. Multiply this by the volume of the solution:

Amount of protein $= (5.35 \times 10^{-5} \text{ M}) \times (50.00 \text{ mL}) \times (1 \text{ L}/1000 \text{ mL}) = 2.67 \times 10^{-6}$ mol.

Molecular weight has units of grams/mole, so simply divide mass by amount:

Molecular Weight of Protein $= (5.00 \text{ g}) \div (2.67 \times 10^{-6} \text{ moles}) = 1.87 \times 10^{6}$ g/mole.

(This is a typical molecular weight for a protein -- these are big molecules!)

Problem: A solution was prepared by dissolving 3.55 g of an unknown non-electrolyte in 7.7 mL of methanol (CH_4O, M.W. = 32.04 g/mol, density = 0.79). The vapor pressure of the resulting solution was 95 Torr. Pure methanol at the same temperature has a vapor pressure of 100 Torr. What is the molecular weight of the unknown compound, assuming it is not volatile?

Solution: Start by rewriting Raoult's Law ($P_{solution} = X_{solvt.} P_{solvt.}$):

$X_{solvent} = P_{solution}/P_{solvent} = 95 \text{ Torr}/100 \text{ Torr} = 0.95$.

Next, calculate the amount of methanol present in the solution:

Amount of Methanol $= (7.7 \text{ mL}) \times (0.79 \text{ g/mL}) \div (32.04 \text{ g/mole}) = 0.19$ moles methanol.

The amount of the unknown compound present can be found by using the definition of mole fraction, substituting the numbers above, and solving:

$X_{solvent} = \frac{\text{moles solvent}}{\text{total moles}} = \frac{\text{moles methanol}}{\text{moles methanol + moles compound}} = \frac{0.19}{0.19 + \text{m.c.}} = 0.95$.

Amount of Compound $= (0.19/0.95) - 0.19 = 0.01$ moles of compound.

The last step is similar to the last step in the first problem:

Molecular Weight of Compound $= 3.55 \text{ g}/0.01 \text{ mole} = 4 \times 10^{2}$ g/mole. (One sig. fig.!)

COLLOIDS:

When you shine a beam of light from a flashlight through the smoke from a campfire, you can follow the path of the flashlight beam with your eyes. This is also possible with the beam of light from the headlights of a car on a foggy night. Normally, however, beams of light passing through the air are invisible. They are made visible in the above cases by the presence of particles (such as the ashes in smoke or the water droplets in fog) which are large enough to reflect the beam of light back to our eyes. (Molecules of air are too small to reflect light.) This phenomenon -- the reflection of light by small, finely-divided particles of one substance dispersed through another substance -- is called the <u>Tyndall effect</u>.

Mixtures which exhibit the Tyndall effect are called <u>colloids</u>. The major difference between colloids and solutions is the size of the particles. True solutions are made up of molecules, ions, and/or particles of similar size. Larger particles, such as the water droplets in fog, are present in colloids. However, colloids cannot be separated by gravity or by filtration, whereas <u>suspensions</u> can. (Examples of suspensions include muddy water or "oil-and-water" salad dressings.)

Smoke and fog are examples of a particular type of colloid called an <u>aerosol</u>. Aerosols are colloids in which one of the two substances present is a gas (air, in the cases above). If both substances present in the colloid are liquids, the colloid is called an <u>emulsion</u>. Milk is an emulsion of butterfat in water. Emulsions do not separate into layers due to the presence of <u>emulsifiers</u>, which are substances that are partially soluble in each substance present in the emulsion. Mayonnaise is an emulsion of oil in water, with egg yolks serving as the emulsifier.[1] Soap is also an emulsifier -- it is partially soluble in both grease and water, and therefore makes grease more water-soluble.

---

1) Cobb, Vicki, <u>Science Experiments You Can Eat</u>, 1972 by J. B. Lippincott Company, New York, p. 36.

THERMODYNAMICS:

If you drop a sugar cube into a glass of water, the sugar cube will eventually dissolve. (Stirring the water or heating it to a higher temperature may speed up the process, but the sugar cube will eventually dissolve even if you don't do these things.) The dissolving of a sugar cube is an example of a spontaneous process -- it happens even without "help" from an outside force (like stirring or heating). On the other hand, some processes are non-spontaneous -- to get them to happen, an outside force must be applied. (For example, you've probably noticed that your room doesn't clean itself up -- you have to "help" it!)

One way to predict whether a process will be spontaneous or not is to determine the amount of energy present in the system under study before and after the process occurs. Any system will tend to undergo a process which lowers its energy spontaneously. Therefore, it's not too surprising that most exothermic reactions tend to occur spontaneously -- an exothermic reaction gives off energy, leaving less energy present in the system at the end of the reaction than it had at the beginning of the reaction. Similarly, most endothermic reactions tend not to occur spontaneously, since these reactions increase the energy of the system.

The word "thermodynamics" comes from "thermo-", meaning "heat", and "dynamic", meaning "changing". The study of thermodynamics is thus the study of the changes in the amount of energy (usually measured as heat) present in systems as they undergo various processes and/or reactions. Thermodynamics also addresses the question of the spontaneity of a process, and allows one to determine the conditions necessary to make a process or reaction occur. (It should be noted here that when a chemist or physicist refers to a process as spontaneous, he or she is referring to the tendency of that process to proceed to completion without adding any outside force. For example, a sugar cube will completely dissolve in water.)

14

THE FIRST LAW OF THERMODYNAMICS:

The three major laws of thermodynamics are known simply as "the first law", "the second law", and "the third law". The First Law of Thermodynamics is also known as the Law of Conservation of Energy: Energy is never created or destroyed, but merely changed from one form to another or transferred from one place to another. (Or, in Einstein's words: "Die Energie der Welt ist Konstant" -- that is, the (total amount of) energy in the universe is a constant.)

Energy is transferred from one place to another in one of two basic forms: heat, or work. This should come as no great surprise if you consider how an engine runs (the engine in a car, for example). The running of the engine does work, which enables the car to move. However, as the engine runs, it gets warmer. Both the work done by the engine and the heat the engine gives off are forms of energy which were originally stored in the engine itself.

Work is the action of one force against an opposing force. (Or, in the words of Willamette University Physics Professor Dr. Maurice Stewart, "Work is what you have to do to get something to happen that won't happen by itself.") You are probably familiar with many forms of work, such as mechanical work (lifting a heavy object exerts force against the force of gravity) and electrical work (pushing a current of electricity through a wire exerts force against the wire's electrical resistance). One form of work which may not be as familiar is pressure-volume work, which is the work done when a gas expands, contracts, or is otherwise re-shaped. Some pressure-volume work is useful work -- for example, the pistons in an internal combustion engine compress gasoline vapors before they are combusted. When these vapors combust, the expanding gases move the pistons, doing work on them and making the engine run. However, much pressure-volume work is relatively useless. As an example, you do P-V work just by walking across a room, thus moving air molecules.

THE SECOND LAW OF THERMODYNAMICS:

When the Exxon Valdez ran aground in March of 1989, the oil it spilled blackened hundreds of miles of Alaskan coastline. This was predictable -- spilled liquids tend to "spread out", occupying as much space as possible. Similarly, we can predict that a scrambled egg cannot be "unscrambled", and that an ice cube will first melt and then evaporate on a warm summer's day.

All of the above phenomena are related, in the sense that they are all examples of how _entropy_ works. _Entropy_ is the term used by scientists to describe the relative "disorder" or "randomness" of a system. For example, an ice cube is a highly _ordered_ system, with the water molecules arranged neatly into a crystalline lattice. Liquid water is less ordered, and water vapor is highly _disordered_ -- the water molecules are free to move in all directions. In going from a solid to a liquid to a gas, the "disorder" of the water -- that is, its _entropy_ -- _increases_.

Any process which increases the entropy of a system will tend to occur spontaneously. This explains why the melting of ice and the evaporation of water both tend to occur spontaneously. These are both _endothermic_ processes, but they occur spontaneously because the _entropy_ of the system increases in each case. This principle also explains why scrambled eggs don't "unscramble" -- to do so would produce a _more ordered_ system, with _less_ entropy, which is _not_ a favorable event.

The Second Law of Thermodynamics generalizes the above principle to cover all processes and reactions. The _Second Law of Thermodynamics_ states that the total entropy of the universe _increases_ whenever a spontaneous change occurs. However, since _everything_ that happens in the universe is either a spontaneous change or the result of a previous spontaneous change, the Second Law has also been stated in this way: "Die Entropie der Welt strebt einen Maximum zu." (Einstein's words for "the (total) entropy of the universe strives to attain a maximum value.")

16

THE THIRD LAW OF THERMODYNAMICS:

The molecules of a gas move around rapidly -- they have lots of kinetic energy. As the gas condenses to form a liquid, the molecules move less rapidly and become more ordered. As the liquid crystallizes to form a solid, the molecules are arranged into the pattern of a crystalline lattice (a highly ordered state), which restricts their movements still further. However, even in a crystalline lattice, molecules have <u>some</u> kinetic energy -- they move around slightly in the lattice. As the temperature gets colder, then, the molecules have less and less kinetic energy. The <u>entropy</u> of the system also decreases. At the absolute zero of temperature, the kinetic energy of the molecules and the entropy of the system would both equal <u>zero</u>. However, this phenomenon has <u>never</u> been observed! Very cold temperatures have been obtained (about 1×10^{-6} K), but it is <u>impossible</u> to make molecules stand still.

The Third Law of Thermodynamics summarizes the above observations. <u>The Third Law of Thermodynamics</u> states that the entropy of a pure, perfect crystalline solid is zero at a temperature of absolute zero. However, since this phenomenon has <u>never</u> been observed, a corollary of the Third Law is that it is not possible to obtain a temperature of absolute zero. (The Third Law has sometimes been called the Law of the Non-Availability of Absolute Zero.)

For many years, people have tried to meet the world's energy needs by building machines that supposedly produce more energy than they consume. However, the laws of thermodynamics show that all such efforts must fail. The First Law says that it isn't possible to produce <u>more</u> energy than is consumed -- at best, the energy produced would be <u>equal</u> to the energy consumed. However, the Second Law says that some energy will be lost as the machine runs, since the <u>entropy</u> of the system must increase. The Third Law says that the only way to avoid this problem is to run the machine at absolute zero, which is a physical impossibility.

FREE ENERGY:

Some of the total energy of a reaction will be unavailable for doing useful work, since some of the total energy of the reaction is lost as the entropy of the system increases. How much of the total energy of a reaction is available to do useful work? The answer to this question is provided by a thermodynamic equation invented by the American scientist J. Willard Gibbs. The equation is: $G = H - TS$, where T is the Kelvin temperature, H refers to <u>enthalpy</u>, S refers to <u>entropy</u> (Note: be careful when spelling "enthalpy" and "entropy" -- they <u>sound</u> similar, but are in fact quite different!), and G refers to the <u>free energy</u> -- that is, the amount of energy "free" to do useful work. (The "G" comes from <u>G</u>ibbs.)

Since chemists are mostly interested in the <u>change</u> in free energy of a given chemical <u>reaction</u>, a more useful form of the above equation is given below: $\Delta G_r^o = \Delta H_r^o - T\Delta S_r^o$. Here, the subscripts "r" refer to a chemical <u>reaction</u>, and the superscripts "o" refer to the <u>standard conditions</u> (25 ^{o}C, 1.00 atm of pressure) under which these quantities are usually measured.

All of the above quantities -- G, H, T, and S -- are <u>state functions</u>. A change in their value depends only on the initial and final <u>states</u> of the system, and not on the path taken by the system to get from the initial state to the final state. Thus, $\Delta S_r^o = S_{products}^o - S_{reactants}^o$ (S^o is called the <u>standard entropy</u> of a substance), and $\Delta G_r^o = \Delta G_f^o$ (products) $- \Delta G_f^o$ (reactants) for a reaction. (ΔG_f^o is called the <u>standard free energy of formation</u> of a substance, and refers to the free energy change of the reaction in which <u>one mole</u> of a substance is formed from its elements. The standard entropy is also a "per mole" quantity.)

One of the most useful features of the free energy is that it allows the prediction of the <u>spontaneity</u> of a reaction. <u>Spontaneous</u> reactions have values of ΔG_r^o which are <u>negative</u>; <u>non-spontaneous</u> reactions have <u>positive</u> ΔG_r^o values.

CALCULATING FREE ENERGIES OF REACTIONS:

Problem: Ethanol (C_2H_6O) has been proposed as an altenative fuel for automobiles. What is the maximum amount of useful work available from combustion of one mole of ethanol? The reaction and ΔG_f^0 values are given below:

$$C_2H_6O_{(1)} + 3\ O_{2\ (g)} \longrightarrow 2\ CO_{2\ (g)} + 3\ H_2O_{(g)}$$

ΔG_f^0:   -41.77   0.00   -94.26   -54.64 (all kcal/mole)[1]

Solution: The maximum amount of useful work available is simply ΔG_r^0, which can be found using the equation: $\Delta G_r^0 = \Delta G_f^0$ (products) $- \Delta G_f^0$ (reactants).

$$\Delta G_r^0 = (2\ \text{moles } CO_2)(-94.26\ \text{kcal/mole}) + (3\ \text{moles } H_2O)(-54.64\ \text{kcal/mol})$$
$$- [(1\ \text{mole } C_2H_6O)(-41.77\ \text{kcal/mole}) + (3\ \text{moles } O_2)(0.00)]$$

$\Delta G_r^0 = $ -310.67 kcal/mole of C_2H_6O. (Reaction is spontaneous.)

Problem: Consider the reaction: $HCl_{(g)} + NH_{3\ (g)} \longrightarrow NH_4Cl_{(s)}$. For this reaction, $\Delta H_r^0 = -42.28$ kcal/mole and $\Delta S_r^0 = -68.03$ cal/mole K.[1] Is this reaction spontaneous at 300 K ? Is this reaction spontaneous at 900 K ?

Solution: Note that both the enthalpy change and the entropy change are negative. Exothermic reactions are usually spontaneous, but reactions in which the entropy decreases are usually non-spontaneous. Which factor will predominate? The answer is, it depends on the temperature. Use $\Delta G_r^0 = \Delta H_r^0 - T\Delta S_r^0$ to find out!

At 300 K: $\Delta G_r^0 = (-42.28\ \text{kcal/mole}) - \dfrac{(300\ \text{K})(-68.03\ \text{cal/mole K})}{(1000\ \text{cal/kcal})}$

$\Delta G_r^0 = $ -21.87 kcal/mole. (Reaction is spontaneous.)

At 900 K: $\Delta G_r^0 = (-42.28\ \text{kcal/mole}) - \dfrac{(900\ \text{K})(-68.03\ \text{cal/mole K})}{(1000\ \text{cal/kcal})}$

$\Delta G_r^0 = $ +18.95 kcal/mole. (Reaction is non-spontaneous.)

Notice that the reaction is spontaneous at room temperature, but not at the higher temperature! (The reverse reaction is spontaneous at the higher temperature.)

---

1) CRC Handbook of Chemistry and Physics, 56th Edition, 1975-1976.

FREE ENERGY AND DYNAMIC EQUILIBRIUM:

If the value of ΔG_r^0 for a reaction is negative, the reaction occurs spontaneously. If the value of ΔG_r^0 for a reaction is positive, the reaction will not occur spontaneously. However, since <u>reversing</u> the reaction simply changes the sign of ΔG_r^0, a positive value of ΔG_r^0 implies that the <u>reverse</u> reaction occurs spontaneously. When ΔG_r^0 has a value of <u>zero</u>, <u>both</u> the forward reaction and the reverse reaction occur simultaneously -- that is, the reaction is in a state of <u>dynamic equilibrium</u>.

<u>Problem</u>: Consider the reaction: $HCl_{(g)} + NH_{3(g)} \longrightarrow NH_4Cl_{(s)}$ For this reaction, $\Delta H_r^0 = -42.28$ kcal/mole and $\Delta S_r^0 = -68.03$ cal/mole K.[1] At what temperature will this reaction come to a state of dynamic equilibrium?

<u>Solution</u>: The equation $\Delta G_r^0 = \Delta H_r^0 - T\Delta S_r^0$ can be used to solve this problem. Since $\Delta G_r^0 = 0$ at equilibrium, $\Delta H_r^0 - T\Delta S_r^0 = 0$ at equilibrium. This implies that $\Delta H_r^0 = T\Delta S_r^0$ at equilibrium, and therefore the temperature (T) at equilibrium is given by: $T = \Delta H_r^0 / \Delta S_r^0$. Substituting the above values, we get:

$$T = \Delta H_r^0 / \Delta S_r^0 = \frac{-42.28 \text{ kcal/mol}}{-68.03 \text{ cal/mole K}} \times \frac{1000 \text{ cal}}{1 \text{ kcal}} = \underline{621.5 \text{ K}}.$$

It should be noted that a system which is at a state of dynamic equilibrium is not capable of doing any useful work, since the value of ΔG_r^0 is a measure of the amount of useful work that a system can do and ΔG_r^0 has a value of <u>zero</u> for a system at equilibrium. You may have experienced this phenomenon before. For example, an automobile battery runs because of chemical reactions which occur inside it. A fully charged automobile battery is <u>not</u> at equilibrium, but as the chemical reactions which power the battery occur, the battery discharges and comes to equilibrium as it does. When the battery finally comes to equilibrium, it is called a "dead battery", and it must be recharged in order to do useful work again.

---

1) CRC Handbook of Chemistry and Physics, 56th Edition, 1975-1976.

CHEMICAL KINETICS:

Some reactions take place at a faster rate than others. For example, the burning of coal and the rusting of iron are basically the same reaction -- oxidation -- taking place at different speeds. The branch of chemistry that deals with the rates at which chemical reactions occur is called chemical kinetics. (It should be noted that while chemical thermodynamics can be used to predict whether or not a reaction can occur, it says nothing about how rapidly the reaction occurs. Some reactions which are spontaneous are too slow to have any practical use, so the study of chemical kinetics is important for "real-world" chemical processes.)

Many factors influence the rates at which reactions occur. Some things are just more reactive than others -- for example, carbon (coal) reacts faster with oxygen than iron does. However, reactions don't occur at all until the reactants are brought into contact with each other. The greater the extent to which the reactants are mixed, the faster the reaction. (For example, homogeneous reactions usually occur more rapidly than heterogeneous reactions, since the reactants in a heterogeneous reaction come together only at the interface of the two phases in the reaction mixture.) Reactions occur when molecules collide, so the more molecules of reactants present per unit volume in a reaction mixture, the greater the chance for reactive collisions to occur. Thus, increasing the concentrations of the reacting species generally increases the rate of the reaction. However, some of the many collisions between molecules do not lead to a reaction occurring, due to insufficient force of collision between the reacting molecules or to an incorrect orientation of the colliding molecules. Increasing the temperature of a reaction provides more energy for the colliding molecules and increases the probability of collisions occurring with the proper orientation. Lastly, catalysts are substances which, when added to a reaction, affect the reaction's rate without being consumed.

RATE EQUATIONS:

The rate at which a chemical reaction proceeds can be determined by measuring either the rate at which the products of the reaction are formed or the rate at which the reactants are consumed. In either case, the measurement of the concentrations of various substances present in the reaction mixture is done at regular intervals. The rate of the reaction is simply the rate at which the concentration (of reactants or products) changes as a function of time.

As the reaction proceeds, the concentration of the reactants decreases as the reactants are consumed. As a result, the chance of collisions between the molecules of the reactants decreases as time elapses. Hence, the rate of the reaction also decreases. The significance of this is that it shows that the rate of a reaction is proportional to the concentrations of the reactants. For the "generic" reaction A + B $\longrightarrow$ C, this can be represented by the rate equation: Rate = $k[A]^a[B]^b$, where [A] and [B] are the concentrations (in moles/liter) of A and B respectively, a and b are the exponents to which those concentrations are raised, and k is the proportionality constant, usually called the rate constant.

The exponents a and b and the rate constant k can only be determined by doing experiments -- they have nothing to do with the stoichiometry of the reaction. Doing this kind of experiment can be challenging, however, since the rate of the reaction decreases as time elapses. One way to get around this problem is to measure the rate of the reaction as soon as the reactants are combined, before the rate has a chance to begin decreasing. This is known as the method of initial rates, since the rate is measured during the initial few minutes of the reaction. Measuring the initial rate for several different concentrations of reactants allows for the determination of the exponents in the rate equation. Once the exponents are known, the rate constant can be found by simple arithmetic, using the rate data.

THE METHOD OF INITIAL RATES:

Problem: The rate equation for the reaction: $C_5H_{10} + HCl \longrightarrow C_5H_{11}Cl$ can be written as: Rate = $k[C_5H_{10}]^a[HCl]^b$. Find the values of the exponents a and b and the rate constant k. The data from several experiments are given below.[1]

Expt.	$[C_5H_{10}]$, M	[HCl], M	Initial Rate, M/sec
#1	0.00920	0.0322	1.76×10^{-5}
#2	0.0203	0.0329	3.96×10^{-5}
#3	0.0618	0.0329	1.20×10^{-4}
#4	0.0877	0.0153	3.92×10^{-5}
#5	0.0872	0.0271	1.22×10^{-4}
#6	0.0918	0.00352	2.07×10^{-6}

Solution: Compare experiments #1 and #2: [HCl] is (roughly) constant, but $[C_5H_{10}]$ goes up by a factor of 2.2. The rate also goes up by a factor of 2.2. In experiments #2 and #3, [HCl] is again constant, but $[C_5H_{10}]$ and the initial rate both go up by a factor of 3.0. Since whatever happens to $[C_5H_{10}]$ happens to the rate, a = 1. (Another way: $(2.2)^a = 2.2$. $(3.0)^a = 3.0$. Therefore, a = 1.)

Now compare experiments #4 and #5: $[C_5H_{10}]$ is (roughly) constant, but [HCl] goes up by a factor of 1.77. The initial rate goes up by a factor of 3.11 -- roughly the square of 1.77. In experiments #4 and #6, $[C_5H_{10}]$ is again (roughly) constant, but [HCl] goes down by a factor of 4.35. The initial rate goes down by a factor of 18.9 -- the square of 4.35. Whatever happens to [HCl] happens squared to the rate, so b = 2. (Another way: $(1.77)^b = 3.11$. $(4.35)^b = 18.9$. Thus, b = 2.)

Now that we know a and b, we can find k using data from experiment #1:

$$k = Rate/[C_5H_{10}][HCl]^2 = (1.76 \times 10^{-5} \text{ M sec}^{-1}) \div [(.00920 \text{ M})(.0322 \text{ M})^2]$$

$$k = 1.85 \text{ M}^{-2} \text{ sec}^{-1}.$$ (Read as "1.85 per molar squared per second".)

---

1) Data taken from Y. Pocker, K. D. Stevens, and J. J. Champoux, _Journal of the American Chemical Society_, vol. 91 (1969), p. 4199.

ORDER OF A REACTION:

The <u>order</u> of a reaction is simply the sum of the exponents in the rate equation for that reaction. Thus, the reaction C_5H_{10} + HCl $\longrightarrow$ $C_5H_{11}Cl$ is a <u>third-order reaction</u>, since its rate equation is: Rate = $k[C_5H_{10}]^1[HCl]^2$. (This reaction is said to be first-order with respect to C_5H_{10}, second-order with respect to HCl, and third-order overall, since 1 + 2 = 3.)

<u>First-order reactions</u> are those whose rate equations have the general form: Rate = $k[R]^1$, where [R] refers to the molar concentration of the reactant. It is possible to use integral calculus to convert this rate equation into another equation which more clearly shows how the concentration of the reactant in a first-order reaction changes as time elapses. We will not go through the calculus here (although readers who understand integral calculus are invited to try to obtain the equation immediately below from the rate equation above!); rather, we will simply show the equation that results. The new equation is: $kt = \ln([R]_i/[R]_f)$, where k is the <u>first-order rate constant</u> for the reaction, t is the amount of <u>time</u> that has elapsed, $[R]_i$ is the <u>initial</u> concentration of the reactant, and $[R]_f$ is the <u>final</u> concentration of the reactant -- that is, the concentration of the reactant after the time represented by t has elapsed. (The symbol "ln" stands for the <u>natural logarithm</u> of the quantity inside the parentheses. Refer to the appendix if you need to refresh your memory (or learn for the first time!) about logarithms.)

<u>Second-order reactions</u> are those whose rate equations have the general form: Rate = $k[R]^2$. Again, integral calculus may be used to convert this equation into the form which more clearly shows how the concentration of the reactant varies as a function of time. Again, we will not do the calculus here, but simply show the new equation for a second-order reaction: $kt = (1/[R]_f) - (1/[R]_i)$, where k is the <u>second-order rate constant</u>, and the other symbols are the same as before.

USING THE CONCENTRATION-TIME RELATIONSHIPS:

Problem: The reaction $C_4H_9Br + OH^- \longrightarrow C_4H_9OH + Br^-$ is a first-order reaction. The rate equation is: Rate = $k[C_4H_9Br]$, with the rate constant k = 0.010 sec^{-1}.[1] If the initial concentration of C_4H_9Br is 0.100 M, how long will it take for 99.0% of the reactant to be consumed?

Solution: If 99.0% of the reactant is consumed, 1.0% of the reactant is not consumed. Hence, the final concentration of C_4H_9Br = 1.0% of 0.100 M = 0.00100 M. Now, simply use the concentration-time equation for first-order cases:

$$kt = \ln([R]_i/[R]_f) = 2.303 \log ([C_4H_9Br]_i/[C_4H_9Br]_f) \quad \text{(see appendix!)}$$
$$(0.010 \text{ sec}^{-1})t = 2.303 \log (0.100 \text{ M}/0.00100 \text{ M}) = 2.303 \log(1.00 \times 10^2)$$
$$(0.010 \text{ sec}^{-1})t = 2.303 (2.000) = 4.606 \quad \text{(see appendix!)}$$
$$t = (4.606) \div (0.010 \text{ sec}^{-1}) = \underline{4.6 \times 10^2 \text{ seconds}}. \quad \text{(Two sig. figs.!)}$$

Problem: The reaction $CH_3Br + OH^- \longrightarrow CH_3OH + Br^-$ is a second-order reaction. The rate equation is: Rate = $k[CH_3Br][OH^-]$, with the rate constant k = 0.0214 $M^{-1} sec^{-1}$.[1] If the initial concentrations of both CH_3Br and OH^- are 0.100 M, what will be the concentrations of the reactants after 1.00 hour?

Solution: Even though this rate equation has a slightly different form from the second-order rate equation discussed previously, the concentration-time relationship can be used as long as the concentrations of the two reactants are equal and the stoichiometry between them is 1:1, as is the case here. Hence:

$$kt = (1/[R]_f) - (1/[R]_i) = (1/[CH_3Br]_f) - (1/[CH_3Br]_i)$$
$$(0.0214 \text{ M}^{-1} \text{ sec}^{-1})(3600 \text{ sec}) = (1/[CH_3Br]_f) - (1/0.100 \text{ M})$$
$$1/[CH_3Br]_f = (77.0 \text{ M}^{-1}) + (10.0 \text{ M}^{-1}) = 87.0 \text{ M}^{-1}$$
$$[CH_3Br]_f = 1/(87.0 \text{ M}^{-1}) = \underline{0.0115 \text{ M}} = [OH^-] \text{ also.} \quad \text{(Three sig. figs.!)}$$

These are examples of "S_N1" and "S_N2" reactions typical of organic molecules.

---

1) Rate constants obtained from R. T. Morrison and R. N. Boyd, Organic Chemistry, Fifth Edition, 1987 by Allyn and Bacon, Inc., page 181.

HALF-LIFE OF A REACTION:

You've probably heard the term "half-life" before in connection with radioactive decay. For example, the radioisotope radon-222 undergoes radioactive decay with a half-life of about four days. This means that <u>half</u> of any given sample of radon-222 will undergo radioactive decay over a four-day period of time. In general, the <u>half-life of a reaction</u> is the amount of time necessary for the concentration of a reactant to be reduced to one-half of its initial value. (In equation form, $[R]_f = [R]_i/2$ if $t = t_{1/2}$ ("$t_{1/2}$" is the symbol for the half-life).)

Given the above information, it is possible to derive expressions which relate the half-life of a reaction to the reaction's rate constant, using the time-concentration relationships. For example, for a <u>first-order reaction</u>:

$$kt = \ln([R]_i/[R]_f) \qquad \text{but if } t = t_{1/2}, [R]_f = [R]_i/2. \text{ Therefore:}$$

$$kt_{1/2} = \ln([R]_i/([R]_i/2)) = \ln(2[R]_i/[R]_i) = \ln 2 = 0.693.$$

Thus, $t_{1/2} = \underline{0.693/k}$ for any <u>first-order</u> reaction.

Notice that the half-life for a first-order reaction is a <u>constant</u> -- it depends only on the rate constant for the reaction, not on the reactant's concentration. This is one reason it is useful in discussing radioactive decay processes, <u>all</u> of which obey first-order kinetics. For a <u>second-order reaction</u>:

$$kt = (1/[R]_f) - (1/[R]_i) \qquad \text{but if } t = t_{1/2}, [R]_f = [R]_i/2. \text{ Thus:}$$

$$kt_{1/2} = (1/([R]_i/2) - (1/[R]_i) = (2/[R]_i) - (1/[R]_i) = 1/[R]_i.$$

Therefore, $t_{1/2} = \underline{1/k[R]_i}$ for any <u>second-order</u> reaction.

<u>Problem</u>: Calculate the half-lives of (a) a first-order reaction with a rate constant of 0.010 sec^{-1}, (b) a second-order reaction with a rate constant of 0.0214 M^{-1} sec^{-1} and an initial reactant concentration of 0.100 M.

<u>Solution</u>: (a) $t_{1/2} = 0.693/k = 0.693/(0.010 \text{ sec}^{-1}) = \underline{69 \text{ seconds}}$.

(b) $t_{1/2} = 1/k[R]_i = 1/[(0.0214 \text{ M}^{-1} \text{ sec}^{-1})(0.100 \text{ M})] = \underline{467 \text{ seconds}}$.

ENERGY OF ACTIVATION:

The burning of wood is simply the reaction of the wood with oxygen in the air. Trees are surrounded by air, but they don't burst into flame unless the temperature around them is increased. For wood to burn, oxygen molecules must collide with the molecules in the wood. These collisions must be sufficiently forceful (that is, they must have enough energy) and must have the molecules in the correct orientation in order for covalent bonds to be broken and new covalent bonds to be formed. The minimum energy that a collision between molecules must possess in order to produce a reaction is called the energy of activation of that reaction.

At low temperatures, molecules move slowly. Collisions between slowly-moving molecules don't have enough energy to react, and the molecules just bounce off each other. At high temperatures, molecules move more rapidly -- that is, they have more kinetic energy. When two rapidly-moving molecules collide, their kinetic energy is transformed into potential energy at the instant the collision occurs. If the potential energy of the molecules at that instant is greater than the energy of activation, the reaction proceeds to give the products by the breaking and forming of covalent bonds. For such a collision, the instant of collision (where the molecules are "stuck together" and the bond-making and bond-breaking processes are in progress) is called the activated complex or the transition state.

In general, increasing the temperature of a reaction by about $10\ ^{O}C$ increases the rate of the reaction by a factor of about three. A more precise correlation between the rate of a reaction, its temperature, and its energy of activation is given by the Arrhenius equation: $\ln k = \ln A - E_a/RT$. Here, k is the rate constant, E_a is the energy of activation, T is the Kelvin temperature, R is the gas constant expressed in energy units (1.987 cal/K mole or 8.314 J/K mole), and A is a "frequency factor" which represents how frequently molecules collide.

USING THE ARRHENIUS EQUATION:

One of the advantages of the Arrhenius equation is that it allows for the determination of the energy of activation of a reaction. This is due to the similarity of the Arrhenius equation ($\ln k = \ln A - E_a/RT$) to the equation of a straight line ($Y = MX + B$). To find the energy of activation of a reaction, the reaction is allowed to take place at several different temperatures, and the rate constant of the reaction is determined for each temperature. A graph is then made, with "$\ln k$" on the Y-axis and "$1/T$" on the X-axis. The data points are plotted, and the straight line which results has a slope (M) equal to $-E_a/R$.

Problem: The following data were obtained for a first-order reaction.[1] Calculate the energy of activation for this reaction.

Expt.	k, sec^{-1}	T, ^{o}C
#1	1.36×10^{-7}	25.0
#2	2.72×10^{-6}	50.1

Solution: First, find "$\ln k$" and "$1/T$" for both experiments:

$\ln k_1 = \ln (1.36 \times 10^{-7}) = \underline{-15.81}$. $1/T_1 = 1/(25.0 + 273) = \underline{0.00336 \ K^{-1}}$.

$\ln k_2 = \ln (2.72 \times 10^{-6}) = \underline{-12.81}$. $1/T_2 = 1/(50.1 + 273) = \underline{0.00310 \ K^{-1}}$.

Now, since only two data points are available, it's easier to find the "slope of the line" by simply finding the quantity "$(Y_2 - Y_1)/(X_2 - X_1)$" rather than plot the line. Substituting "$\ln k$" for "Y" and "$1/T$" for "X", we get:

"Slope" = $(\ln k_2 - \ln k_1)/(1/T_2 - 1/T_1)$

"Slope" = $[(-15.81) - (-12.81)]/[(0.00336 \ K^{-1}) - (0.00310 \ K^{-1})]$

"Slope" = $(-3.00)/(0.00026 \ K^{-1}) = \underline{-1.2 \times 10^4 \ K}$.

Finally, since "Slope" = $-E_a/R$, finding E_a is straightforward:

E_a = $-R \times$ "Slope" = $(-1.987 \ cal/K \ mole)(-1.2 \times 10^4 \ K) = \underline{2.4 \times 10^4 \ cal/mole}$.

---

1) Data taken from A. Streitwieser, Jr. et al., Journal of the American Chemical Society, vol. 92 (1970), p.5141.

MECHANISM OF A REACTION:

Some reactions (such as the "S_N2" reaction: $CH_3Br + OH^- \longrightarrow$ $CH_3OH + Br^-$) occur in a single step -- the reactants collide, and the products are formed. Other reactions (such as the "S_N1" reaction: $C_4H_9Br + OH^- \longrightarrow$ $C_4H_9OH + Br^-$) occur in a series of steps, as shown in the sequence below:

Step #1: $C_4H_9Br \longrightarrow C_4H_9^+ + Br^-$

Step #2: $C_4H_9^+ + OH^- \longrightarrow C_4H_9OH$

A sequence of individual "steps" (or elementary processes, as the "steps" are often called) which describes the way in which a reaction takes place on the molecular level is called the mechanism of the reaction.

It is not possible to prove that a reaction proceeds by a particular mechanism, since we can't "see" molecules as they react. However, support for individual mechanisms can be found in the results of kinetics experiments. For example, consider the reactions above. The "S_N2" reaction has the following rate equation: Rate = $k[CH_3Br][OH^-]$. On the other hand, the "S_N1" reaction has this rate equation: Rate = $k[C_4H_9Br]$. Notice that the rate of the "S_N2" reaction depends on the concentration of OH^-, but the rate of the "S_N1" reaction does not. This is due to the fact that the "S_N1" reaction involves a two-step mechanism, in which the first step (above) occurs much more slowly than the second step (above). The slowest step in a reaction mechanism is called the rate-determining step, since the rate at which this step occurs determines the rate at which the entire reaction occurs. (As an analogy, consider several workers on an assembly line. No matter how fast the other workers work, the amount of work done by the slowest worker determines the rate at which the finished product is produced.) Since OH^- is not involved in the rate-determining step, the rate of the "S_N1" reaction does not depend on $[OH^-]$. Hence, Rate = $k[C_4H_9Br]$, as confirmed by kinetics experiments.

FREE-RADICAL "CHAIN" REACTIONS:

Ozone (O_3) is a form of oxygen which is formed in the upper atmosphere by the breakdown of O_2 into individual atoms of oxygen. These individual oxygen atoms then combine with molecules of O_2 to form molecules of O_3, as shown below:

Step #1: O_2 + energy $\longrightarrow$ 2 O

Step #2: $O + O_2 \longrightarrow O_3$

Since atmospheric ozone absorbs harmful ultraviolet radiation from the sun, thereby protecting us from skin cancer, it is important that the earth's ozone layer remain intact. Thus, there has been a growing concern since 1974 about the destruction of atmospheric ozone by a class of molecules called chlorofluorocarbons (or CFC's). CFC's are used in refrigerators, air conditioners, and aerosol spray cans. The problem with CFC's is that they tend to form free radicals upon exposure to ultraviolet light. Free radicals are atoms or groups of atoms which have at least one unpaired electron. Free radicals tend to be very reactive, since the unpaired electron can lower its energy by finding another electron and forming an electron pair. The sequence below shows how a typical CFC destroys atmospheric ozone:[1]

Step #1: $CFCl_3$ + U.V. light $\longrightarrow$ $CFCl_2$ + :C̈l· (initiation)

Step #2: :C̈l· + O_3 $\longrightarrow$ O_2 + :C̈l-Ö· (propagation)

Step #3: :C̈l-Ö· + O $\longrightarrow$ O_2 + :C̈l· (propagation)

This process is called a free-radical chain reaction, since chlorine free radicals are formed in Step #1 (the initiation step) and then react with ozone in Step #2 and Step #3 (the chain propagation steps). Notice that at the end of Step #3, the chlorine radical generated in Step #1 has been re-generated, and is thus available to destroy another ozone molecule. Since Steps #2 and #3 are repeated many times (hence the "chain" reaction), one CFC molecule can destroy 1000 molecules of O_3.

---

1) TIME Magazine, October 19, 1987 issue.

HOW CATALYSTS WORK:

Catalysts are substances which affect the rate of chemical reactions without being consumed themselves during the course of the reaction. Usually, the effect of a catalyst is to increase the rate of a reaction. This happens because the presence of the catalyst provides an alternative mechanism by which a reaction can proceed. The alternative mechanism usually has a lower energy of activation than the uncatalyzed mechanism. Therefore, the reaction will proceed at a faster rate, since a greater percentage of the collisions between molecules will have a sufficient amount of energy to produce a reaction and lead to formation of product.

There are two main types of catalysts: homogeneous and heterogeneous. Homogeneous catalysts are contained in the same phase as the rest of the reaction mixture during the course of the reaction. An example is sulfuric acid, which catalyzes the reaction between acetic acid and amyl alcohol to form amyl acetate (pear oil). Sulfuric acid, acetic acid, and amyl alcohol form a homogeneous mixture, and the purpose of the sulfuric acid is to react with the acetic acid and make it more susceptible to attack by molecules of amyl alcohol. If desired, the sulfuric acid can be recovered, apparently unreacted, at the end of the reaction. (Sulfuric acid does react, but is re-generated during the reaction.)

Heterogeneous catalysts occupy a different phase from the reactants during the course of the reaction. Typical heterogeneous catalysts include the transition metals nickel, palladium, and platinum. These catalysts are used in the reaction of hydrogen with organic materials in a process called hydrogenation. You're probably most familiar with this process as a way to change vegetable oils into vegetable shortening -- the only difference is a few hydrogen atoms in each molecule! Molecules of H_2 and vegetable oils become attached to the metal surface, which assists in the bond-making and bond-breaking processes in the reaction.

31

DYNAMIC EQUILIBRIUM AND LeCHATELIER'S PRINCIPLE:

During the course of the "generic" reaction R ⟶ P, the concentration of the reactants gradually <u>decreases</u>, and the concentration of the products slowly <u>increases</u>. Since the concentration of the reactants is decreasing, the <u>rate</u> of the reaction will also decrease. However, since the concentration of the products is <u>increasing</u>, the rate of the <u>reverse</u> reaction (P ⟶ R) will tend to <u>increase</u> if the energy of activation of the reverse reaction is not too large to prevent this. As the rate of the forward reaction (R ⟶ P) decreases and the rate of the reverse reaction increases, a point will be reached at which <u>both</u> reactions are proceeding at the <u>same</u> rate. This situation is called <u>dynamic equilibrium</u> -- the system is at <u>equilibrium</u> (since the concentrations of reactants and products do not change), but individual molecules continue to undergo reactions and are therefore <u>dynamic</u>. ("Dynamic" here means "moving" or "changing".)

Dynamic equilibrium is a relatively <u>low</u> energy state for a system. Indeed, the free energy (ΔG) of a system at equilibrium is <u>zero</u>. Most systems try to minimize their energy content whenever possible. Therefore, dynamic equilibrium is a desirable state for a system. If something happens to a system at equilibrium to disturb the system and remove it from its state of equilibrium, then the system will do whatever it can to return to its equilibrium state. This principle was first stated in 1888 by the French physical chemist Henri Louis LeChatelier, and it has come to be known as <u>LeChatelier's Principle</u>.

For the equilibrium reaction rR ⇌ pP (the double arrows represent equilibrium), the <u>mass action expression</u> (sometimes called the <u>reaction quotient</u>, Q) is written as: $[P]^p/[R]^r$. (Note that the <u>coefficients</u> in the balanced equation equal the <u>exponents</u> in the reaction quotient.) The system is at equilibrium when the reaction quotient equals the <u>equilibrium constant</u> (K_{eq}) for that reaction.

FREE ENERGY AND EQUILIBRIUM CONSTANTS:

Ammonia (NH$_3$) can be formed from nitrogen and hydrogen by the reaction:

$$N_2 \text{ (g)} + 3 H_2 \text{ (g)} \rightleftharpoons 2 NH_3 \text{ (g)}.$$ This reaction is called the Haber Process.

Problem: What is the mass action expression for this equilibrium?

Solution: If you remember that the rule is "products-over-reactants", and that the coefficients in the balanced equation equal the exponents in the mass action expression, this should be easy: $[NH_3]^2/[N_2][H_2]^3$.

Problem: A 1.00-liter flask contains 8.2 moles of NH$_3$, 0.100 moles of N$_2$, and 0.100 moles of H$_2$ when the system has come to equilibrium. Calculate the equilibrium constant for the above reaction.

Solution: Since the volume of the flask is 1.00 liter, the amounts given above are numerically the same as the concentrations of each substance. Thus:

$$[NH_3] = 8.2 \text{ M} \qquad [N_2] = 0.100 \text{ M} \qquad [H_2] = 0.100 \text{ M}$$

$$K_{eq} = [NH_3]^2/[N_2][H_2]^3 = (8.2 \text{ M})^2/(0.100 \text{ M})(0.100 \text{ M})^3$$

$$K_{eq} = 6.7 \times 10^5.[1] \quad (\underline{\text{Units}} \text{ are usually } \underline{\text{omitted}} \text{ for equilibrium constants.})$$

Problem: Calculate ΔG_r^0 for the above reaction, given that ΔG_r^0 and K$_{eq}$ are related by the following equation: $\Delta G_r^0 = -RT \ln K_{eq}$.

Solution: This should be easy if you recall that the "o" in "ΔG_r^0" refers to a temperature of 25.0 ^{0}C, which equals 298 K. Therefore:

$$\Delta G_r^0 = -RT \ln K_{eq} = -(1.987 \text{ cal/K mole})(298 \text{ K}) \ln(6.7 \times 10^5)$$

$$\Delta G_r^0 = -(1.987 \text{ cal/K mole})(298 \text{ K})(13.42) = -7.95 \times 10^3 \text{ cal/mole.}[1]$$

This reaction is spontaneous at 25 ^{0}C. The spontaneity of the reaction is also exemplified by the large value of the equilibrium constant. Large values of K$_{eq}$ imply spontaneous reactions (favoring formation of products); small values of K$_{eq}$ imply non-spontaneous reactions (favoring formation of the reactants).

---

1) CRC Handbook of Chemistry and Physics, 56th Edition, 1975-1976.

USING LeCHATELIER'S PRINCIPLE:

The Haber Process reaction, $N_2 (g) + 3 H_2 (g) \rightleftharpoons 2 NH_3 (g)$, is exothermic (its ΔH_r^0 value is -22.08 kcal/mol[1]). LeChatelier's Principle says that if a system at equilibrium is disturbed, either the forward reaction or the reverse reaction will occur predominantly until the equilibrium is restored.

Problem: If $N_2 (g)$ is added to the equilibrium, what will happen?

Solution: Adding N_2 increases $[N_2]$, which increases the rate of the forward reaction. Thus, the forward reaction will occur -- NH_3 will be formed, and N_2 and H_2 will be consumed -- until equilibrium is restored. (This is sometimes summarized by saying that the equilibrium "shifts to the right", towards products.)

Problem: If $NH_3 (g)$ is added to the equilibrium, what will happen?

Solution: Adding NH_3 increases $[NH_3]$, which increases the rate of the reverse reaction. Thus, the reverse reaction will occur -- NH_3 will be consumed, and N_2 and H_2 will be formed -- until equilibrium is restored. (This is sometimes summarized by saying that the equilibrium "shifts to the left", towards reactants.)

Problem: If heat is added to the equilibrium, what will happen?

Solution: Think of it this way -- for an exothermic reaction, heat is a product. Therefore, the situation is similar to adding NH_3 to the system -- the equilibrium will shift to the left. (If the reaction had been endothermic, adding heat would shift the equilibrium to the right. This is van't Hoff's Law.)

Problem: If the container is compressed, what will happen?

Solution: Compressing the container increases all three concentrations. To restore the equilibrium, the system will shift in the direction which allows it to reduce the total number of gas particles. In this case, the shift is to the right, since four moles of gas are on the left, but only two moles are on the right.

----
1) CRC Handbook of Chemistry and Physics, 56th Edition, 1975-76.

SOLVING EQUILIBRIUM PROBLEMS:

Problem: Consider the equilibrium: $PCl_{5\ (g)} \rightleftharpoons PCl_{3\ (g)} + Cl_{2\ (g)}$.
The equilibrium constant for this reaction is 7.30×10^{-2}. If 1.00 mole of PCl_5 is injected into an empty 1.00-liter container, what will be the concentrations of PCl_5, PCl_3, and Cl_2 when the system has come to equilibrium?[1]

Solution: First, write the equilibrium expression for this reaction:
$$K_{eq} = [PCl_3][Cl_2]/[PCl_5] = 7.30 \times 10^{-2}.$$

Now, during the reaction, some PCl_5 will decompose, forming PCl_3 and Cl_2. What we don't know is how much PCl_5 will react in this way. Hence, call the amount of PCl_5 that reacts "X". From the balanced equation above, if "X" moles of PCl_5 react, "X" moles of PCl_3 and "X" moles of Cl_2 will be formed. These results are summarized in the following table:

	Initial Value	Change During Reaction	Final Value
$[PCl_5]$	1.00 M	-X	(1.00 - X) M
$[PCl_3]$	0.00 M	+X	X M
$[Cl_2]$	0.00 M	+X	X M

Inserting the "Final Value" entries into the equilibrium expression above gives:
$$[PCl_3][Cl_2]/[PCl_5] = 7.30 \times 10^{-2} = (X)(X)/(1.00 - X) = X^2/(1.00 - X).$$

This equation can be rewritten in the form of a quadratic equation, as shown below:
$$X^2 = (7.30 \times 10^{-2})(1.00 - X) = 7.30 \times 10^{-2} - (7.30 \times 10^{-2})(X)$$
$$X^2 + (7.30 \times 10^{-2})(X) - 7.30 \times 10^{-2} = 0 = aX^2 + bX + c$$

where $a = 1$, $b = 7.30 \times 10^{-2}$, and $c = -7.30 \times 10^{-2}$. Solving the quadratic equation:
$$X = [-b \pm (b^2 - 4ac)^{1/2}]/2a$$
$$X = [-(7.30 \times 10^{-2}) + ((7.30 \times 10^{-2})^2 - (4)(1)(-7.30 \times 10^{-2}))^{1/2}]/(2)(1) = \underline{0.236}.$$

Thus: $[PCl_5] = 1.00 - 0.236 = \underline{0.76\ M}$. $[PCl_3] = \underline{0.236\ M}$. $[Cl_2] = \underline{0.236\ M}$.

---

1) From a similar problem on page 313 in "Chemical Principles and Properties" by M. J. Sienko and R. A. Plane. 1974 by McGraw-Hill, Inc.

AN ALTERNATIVE METHOD FOR SOLVING SOME EQUILIBRIUM PROBLEMS:

Problem: The equilibrium constant is 1.8×10^{-6} for the equilibrium $2 NO_{2 (g)} \rightleftharpoons 2 NO_{(g)} + O_{2 (g)}$. If 1.00 mole of NO_2 is injected into an empty 1.00-liter container, what will be the concentrations of NO_2, NO, and O_2 when the system has come to equilibrium?[1]

Solution: First, write the equilibrium expression for this reaction:
$K_{eq} = [NO]^2[O_2]/[NO_2]^2 = 1.8 \times 10^{-6}$.

During the reaction, if "X" moles of O_2 are formed, "2X" moles of NO will be formed, and "2X" moles of NO_2 will be consumed. See the table below:

	Initial Value	Change During Reaction	Final Value
$[NO_2]$	1.00 M	-2X	(1.00 - 2X) M
$[NO]$	0.00 M	+2X	2X M
$[O_2]$	0.00 M	+X	X M

Inserting the "Final Value" entries into the equilibrium expression above gives:
$$[NO]^2[O_2]/[NO_2]^2 = 1.8 \times 10^{-6} = (2X)^2(X)/(1.00 - 2X)^2 = 4X^3/(1.00-2X)^2.$$
Solving this cubic equation would be quite challenging! However, there's a "short cut" that we can take here because the equilibrium constant is a very small number. This implies that the forward reaction does not proceed very far -- that is, that "X" will also be a very small number. If "X" is very small, then the "2X" in the denominator above can be neglected with respect to the "1.00" from which it is being subtracted. Hence, the denominator is approximately equal to 1.00. This makes solving the above equation much easier, as shown below:
$$4X^3/(1.00)^2 = 1.8 \times 10^{-6} = 4X^3. \quad X^3 = (1.8 \times 10^{-6})/4 = 4.5 \times 10^{-7}.$$
$$X = (4.5 \times 10^{-7})^{1/3} = \underline{7.7 \times 10^{-3}} = [O_2] \text{ at equilibrium.}$$
Thus: $[NO] = (2)(7.7 \times 10^{-3}) = \underline{1.5 \times 10^{-2}}$ M. $[NO_2] = 1.00 - 0.015 = \underline{0.98}$ M.

---

1) From a similar problem on page 473 in "General Chemistry" by R. H. Petrucci. Fourth Edition, 1985 by Macmillan Publishing Company.

HETEROGENEOUS EQUILIBRIA:

A $\underline{heterogeneous\ equilibrium}$ is one in which the species in equilibrium are present in different physical states or phases. An example is the equilibrium $C_{(s)} + O_{2\ (g)} \rightleftharpoons CO_{2\ (g)}$, in which $\underline{solid}$ carbon and two $\underline{gases}$ are all in equilibrium. You might think that the equilibrium expression for this equilibrium would be written as: $K_{eq} = [CO_2]/[C][O_2]$. However, unlike the concentrations of $\underline{gases}$ (which can be compressed) or $\underline{solutions}$ (which have variable compositions), the "concentrations" of $\underline{pure\ solids}$ and $\underline{pure\ liquids}$ are $\underline{constants}$, since they only depend on the density and molecular weight (both constants) of the solid or liquid. Hence, the "concentration" term for a pure solid or pure liquid is incorporated as part of the equilibrium constant, and the equilibrium expressions of heterogeneous equilibria do not include these terms. Hence, the "[C]" term should not appear in the equilibrium expression above, and the correct way to write this equilibrium expression is: $K_{eq} = [CO_2]/[O_2]$. In a similar fashion, the equilibrium expression for the heterogeneous equilibrium $2\ H_{2\ (g)} + O_{2\ (g)} \rightleftharpoons 2\ H_2O_{(1)}$ should be written as: $K_{eq} = 1/[H_2]^2[O_2]$, since $H_2O_{(1)}$ is a pure liquid.

A $\underline{saturated\ solution}$ is prepared by adding solute to a solvent until no more solute will dissolve, and then adding extra solute, which sinks to the bottom of the container. In this system, the undissolved solute is in equilibrium with the dissolved solute. This is therefore a heterogeneous equilibrium. For example, the equilibrium $BaSO_{4\ (s)} \rightleftharpoons Ba^{2+}_{\ (aq)} + SO_4^{2-}_{\ (aq)}$ is present in a saturated solution of barium sulfate. Since $[BaSO_4]$ is a constant, it is omitted from the equilibrium expression, and the correct form of the equilibrium expression is: $K_{eq} = [Ba^{2+}][SO_4^{2-}]$. Since this is simply the $\underline{product}$ of the concentrations of the ions in solution, it is usually called the $\underline{solubility\ product}$, and the equilibrium constant is called the $\underline{solubility\ product\ constant}$, or K_{sp} for short.

Problem: The concentration of Ba^{2+} ions in a saturated solution of $BaSO_4$ is 1.04×10^{-5} M. What is the solubility product constant (K_{sp}) for $BaSO_4$?

Solution: The equilibrium is: $BaSO_{4\ (s)} \rightleftharpoons Ba^{2+}_{\ (aq)} + SO_4^{2-}_{\ (aq)}$. Thus, equal numbers of barium ions and sulfate ions must be present in solution. Therefore, $[SO_4^{2-}] = [Ba^{2+}]$, and the value of K_{sp} is easily calculated, as shown:

$$K_{sp} = [Ba^{2+}][SO_4^{2-}] = (1.04 \times 10^{-5})(1.04 \times 10^{-5}) = \underline{1.08 \times 10^{-10}}.[1]$$

Problem: The K_{sp} value for silver chloride (AgCl) is 1.56×10^{-10}.[1] Calculate $[Ag^+]$ and $[Cl^-]$ in a saturated aqueous solution of AgCl.

Solution: The equilibrium is: $AgCl_{\ (s)} \rightleftharpoons Ag^+_{\ (aq)} + Cl^-_{\ (aq)}$. Call the amount of AgCl which dissolves "X". If "X" moles of AgCl dissolve, "X" moles of Ag^+ ions and "X" moles of Cl^- ions must be formed. See the table below:

	Initial Value	Change During Reaction	Final Value
$[Ag^+]$	0.00 M	+X	X M
$[Cl^-]$	0.00 M	+X	X M

Inserting the "Final Value" entries into the equilibrium expression gives:

$$K_{sp} = 1.56 \times 10^{-10} = [Ag^+][Cl^-] = (X)(X) = X^2.$$
$$X = (1.56 \times 10^{-10})^{1/2} = \underline{1.25 \times 10^{-5}} = [Ag^+] = [Cl^-].$$

Problem: The K_{sp} value for lead iodide (PbI_2) is 1.39×10^{-8}.[1] Calculate $[Pb^{2+}]$ and $[I^-]$ in a saturated aqueous solution of lead iodide.

Solution:

(The equilibrium is:		Initial	Change	Final
	$[Pb^{2+}]$	0.00 M	+X	X M
$PbI_{2(s)} \rightleftharpoons Pb^{2+}_{\ (aq)} + 2\ I^-_{\ (aq)}.)$	$[I^-]$	0.00 M	+2X	2X M

$$K_{sp} = 1.39 \times 10^{-8} = [Pb^{2+}][I^-]^2 = (X)(2X)^2 = 4X^3.$$
$$X = (1.39 \times 10^{-8}/4)^{1/3} = \underline{1.51 \times 10^{-3}} = [Pb^{2+}]. \quad [I^-] = \underline{3.03 \times 10^{-3}}\ M.$$

1) CRC Handbook of Chemistry and Physics, 56th Edition, 1975-1976.

THE COMMON ION EFFECT:

The solubility of an ionic solute in water is <u>decreased</u> if the water already contains one of the ions which is present in the solute. This observation is called the <u>common ion effect</u>, and is illustrated by the following situation:

<u>Problem</u>: The K_{sp} value for lead iodide (PbI_2) is 1.39×10^{-8}.[1] Calculate $[Pb^{2+}]$ in a saturated aqueous solution of lead iodide if enough potassium iodide (KI) is added to make $[I^-] = 0.100$ M.

<u>Solution</u>: The equilibrium is: $PbI_{2\,(s)} \rightleftharpoons Pb^{2+}_{(aq)} + 2\,I^-_{(aq)}$. Call the amount of PbI_2 which dissolves "X". If "X" moles of PbI_2 dissolve, "X" moles of Pb^{2+} ions and "2X" moles of I^- ions will be formed. However, the <u>total</u> concentration of I^- ions in the solution is already known: $[I^-] = 0.100$ M. Thus:

	Initial Value	Change During Reaction	Final Value
$[Pb^{2+}]$	0.00 M	+X	X M
$[I^-]$	0.00 M	+2X	0.100 M

Inserting the "Final Value" entries into the equilibrium expression gives:

$$K_{sp} = 1.39 \times 10^{-8} = [Pb^{2+}][I^-]^2 = (X)(0.100)^2 = (X)(0.0100).$$
$$X = (1.39 \times 10^{-8})/(0.0100) = \underline{1.39 \times 10^{-6}} = [Pb^{2+}].$$

Notice that this concentration is <u>lower</u> than the concentration of Pb^{2+} in a saturated aqueous solution of PbI_2 <u>without</u> added iodide ion ($[Pb^{2+}] = 1.51 \times 10^{-3}$). This should not be surprising, since LeChatelier's Principle predicts that adding I^- ions to the above equilibrium should shift it to the <u>left</u>, <u>lowering</u> $[Pb^{2+}]$.

<u>Problem</u>: Calculate $[I^-]$ in a saturated aqueous solution of lead iodide if enough lead nitrate ($Pb(NO_3)_2$) is added to make $[Pb^{2+}] = 0.100$ M.

<u>Solution</u>: $K_{sp} = 1.39 \times 10^{-8} = [Pb^{2+}][I^-]^2 = (0.100)(X)^2.$

$$X = [(1.39 \times 10^{-8})/(0.100)]^{1/2} = (1.39 \times 10^{-7})^{1/2} = \underline{3.73 \times 10^{-4}} = [I^-].$$

---

1) CRC Handbook of Chemistry and Physics, 56th Edition, 1975-1976.

SOME "REAL WORLD" EFFECTS OF DYNAMIC EQUILIBRIA:

Dynamic equilibria affect us in many ways. Most notable are the many biological and biochemical processes that depend on dynamic equilibria. As an example, consider the following equilibrium:

$$Ca_{10}(PO_4)_6(OH)_2 + 2\ F^- \rightleftharpoons Ca_{10}(PO_4)_6F_2 + 2\ OH^-$$

(hydroxyapatite)   (fluoroapatite)

Hydroxyapatite and fluoroapatite are the mineral forms of calcium found in human bones and teeth. "Fluoroapatite is less easily dissolved by mouth acids than hydroxyapatite and therefore more resistant to decay."[1] Increasing the amount of fluoride ions present in the bloodstream shifts the above equilibrium to the <u>right</u>, favoring the formation of the acid-resistant fluoroapatite. This is one reason why many dentists advise drinking fluoridated water and using toothpaste which contains fluoride ions. The fluoridation of water remains controversial, as some people are adversely affected by an excess of fluoride in their bloodstreams.[1]

Another example involves the tendency of chickens to lay eggs with shells that are too thin to survive handling and processing. This usually occurs in hot weather, when chickens are more likely to pant to rid themselves of excess body heat. "Panting can appreciably lower the level of carbon dioxide in a chicken's blood."[2] This shifts the following equilibrium to the <u>left</u>:

$$H_2O\ (l) + CO_2\ (g) \rightleftharpoons 2\ H^+\ (aq) + CO_3^{2-}\ (aq)$$

"The shift reduces the amount of carbonate (CO_3^{2-}) ion available to the chicken for making eggshells, which are 95% calcium carbonate ($CaCO_3$) in the form of calcite. Poultry farmers could solve the problem by having their flocks drink Perrier water or just plain carbonated water."[2] Experiments support this claim.[3]

1) <u>Chemical & Engineering News</u>, August 1, 1988, page 29.
2) <u>Chemical & Engineering News</u>, March 20, 1989, page 94.
3) Odun, T. W., <u>et al</u>., <u>Poultry Science</u>, vol. 64 (1985), page 594, as cited in <u>Chemical & Engineering News</u>, June 12, 1989, page 48.

THE SELF-IONIZATION OF WATER:

Consider the equilibrium: $H\!:\!\overset{H}{\underset{\cdot\cdot}{\overset{\cdot\cdot}{O}}}\!:$ + $H\!:\!\overset{H}{\underset{\cdot\cdot}{\overset{\cdot\cdot}{O}}}\!:$ ⇌ $H\!:\!\overset{H}{\underset{\cdot\cdot}{\overset{\cdot\cdot}{O}}}\!:\!H$ + $H\!:\!\overset{H}{\underset{\cdot\cdot}{\overset{\cdot\cdot}{O}}}\!:$, which can also be written as: $2 H_2O_{(1)}$ ⇌ $H_3O^+_{(aq)}$ + $OH^-_{(aq)}$. Notice that the water molecule on the right transfers a proton to the water molecule on the left. (A hydrogen ion, H^+, is commonly called a proton, since that's what it is -- a hydrogen atom which has lost its only electron is simply a proton!) This occurs to a small extent in any system which contains water. The above equilibrium is a heterogeneous equilibrium, so the equilibrium expression does not include a $[H_2O]$ term. The equilibrium expression for the above equilibrium is: $[H_3O^+][OH^-] = K_w$, where K_w (the ion product constant for water, hence the "w") has a value of 1.00×10^{-14} at a temperature of 25 °C. The above equilibrium is sometimes called the self-ionization of water, since water molecules form ions when they interact.

The above equilibrium also illustrates several ways of looking at acid-base chemistry. Acids and bases have been defined in several ways, all of which can be useful. The Arrhenius definition describes acids as compounds which form H_3O^+ ions (called hydronium ions) when placed in water, and bases as compounds which form OH^- ions (called hydroxide ions) when placed in water. (Water itself is amphoteric -- that is, able to react as either an acid or a base -- since both H_3O^+ and OH^- ions are formed in the above reaction.) The Brønsted definition describes acids as proton donors, and bases as proton acceptors. In the above reaction, the water molecule on the right "donates" a proton to the water molecule on the left, which "accepts" the proton. (Again, water behaves as either an acid or a base.) The Lewis definition describes acids as electron-pair acceptors, and bases as electron-pair donors. In the above reaction, notice that the water molecule on the left "donates" one of the lone pairs of electrons on the oxygen atom to the proton which "accepts" it, thereby forming a coordinate covalent bond.

ACIDIC, BASIC, AND NEUTRAL SOLUTIONS:

Problem: Calculate $[H_3O^+]$ and $[OH^-]$ in pure water, given the equilibrium: $2 H_2O_{(l)} \rightleftarrows H_3O^+_{(aq)} + OH^-_{(aq)}$. $K_w = 1.00 \times 10^{-14}$.

Solution: Since the reaction of two moles of water molecules produces one mole of hydronium ions and one mole of hydroxide ions, $[H_3O^+] = [OH^-]$. Thus:

$$K_w = [H_3O^+][OH^-] = 1.00 \times 10^{-14} = (X)(X) = X^2.$$

$$X = (1.00 \times 10^{-14})^{1/2} = 1.00 \times 10^{-7} = [H_3O^+] = [OH^-].$$

Pure water is neutral, since $[H_3O^+] = [OH^-]$. In an acidic solution, $[H_3O^+]$ is greater than $[OH^-]$, and in a basic solution, $[OH^-]$ is greater than $[H_3O^+]$.

To avoid the extensive use of scientific notation when discussing the acidity or basicity of aqueous solutions, the terms "pH" and "pOH" are often used. The "p" refers to the negative logarithm of the quantity in question. Therefore, pH = $-\log[H_3O^+]$, and pOH = $-\log[OH^-]$. ("pH" = $-\log[H^+]$ = $-\log[H_3O^+]$.)

Problem: Calculate the pH and the pOH of neutral water.

Solution: In neutral water, $[H_3O^+] = [OH^-] = 1.00 \times 10^{-7}$ M.

Therefore, pH = pOH = $-\log(1.00 \times 10^{-7})$ = 7.000.

Notice that pH + pOH = 7.000 + 7.000 = 14.000. The sum of the pH and the pOH of any aqueous solution will always be equal to 14.000 at 25 °C. (This should not be surprising, since 14.000 is simply "pK_w" -- the negative logarithm of K_w.)

In an acidic solution, the pH is less than 7.000. (This may be a bit confusing, since $[H_3O^+]$ is greater than 1.00×10^{-7} M in an acidic solution, but the negative sign in the definition of pH causes the trend to be reversed.) Since pH + pOH = 14.000, it follows that pOH is greater than 7.000 in an acidic solution. In a basic solution, exactly the opposite is true -- the pH is greater than 7.000, and the pOH is less than 7.000. Remember, though -- regardless of whether the solution is acidic, basic, or neutral, the sum of the pH and the pOH is 14.000.

USING pH AND pOH:

One way in which pH and pOH are typically used is to describe the acidity or basicity of <u>dilute</u> solutions of <u>strong</u> acids and <u>strong</u> bases. A <u>strong</u> acid or base is, by definition, <u>totally</u> dissociated into ions when placed in water. Therefore, for <u>strong</u> acids, [acid] = $[H_3O^+]$, and for <u>strong</u> bases, [base] = $[OH^-]$.

<u>Problem</u>: What is the pH of a 0.0100 M HCl solution?

<u>Solution</u>: HCl is a strong acid, so [HCl] = 1.00×10^{-2} M = $[H_3O^+]$. Therefore, pH = $-\log[H_3O^+]$ = $-\log(1.00 \times 10^{-2})$ = <u>2.000</u>.

<u>Problem</u>: What is the pH of a 2.0×10^{-4} M NaOH solution?

<u>Solution</u>: NaOH dissolves completely in water to give Na^+ and OH^- ions. Therefore, [NaOH] = 2.0×10^{-4} M = $[OH^-]$. Now, using pH + pOH = 14.000, we get:

pOH = $-\log[OH^-]$ = $-\log(2.0 \times 10^{-4})$ = 3.70.

pH = 14.000 - pH = 14.000 - 3.70 = <u>10.30</u>.

<u>Problem</u>: "Normally, rain has a pH of 5.6, but over much of the Northeast it's an acidic 4.3."[1] How many times more acidic is Northeastern rain than normal rain? (Rain which has a pH value of 5.0 or less is called "acid rain".)

<u>Solution</u>: One way to approach this problem is to convert the pH values above into $[H_3O^+]$ values and then divide the two $[H_3O^+]$ values. Hence:

Normal rain: $[H_3O^+]$ = 10^{-pH} = $10^{-5.6}$ = <u>2.5×10^{-6} M</u>.

Northeastern rain: $[H_3O^+]$ = 10^{-pH} = $10^{-4.3}$ = <u>5.0×10^{-5} M</u>.

$(5.0 \times 10^{-5}$ M$)/(2.5 \times 10^{-6}$ M$)$ = <u>20</u>. Therefore, Northeastern rain is <u>twenty times more acidic than normal rain</u>. This points out an important feature of using pH and pOH: due to the use of the logarithmic function, a change of <u>1.0 unit</u> of pH or pOH corresponds to a <u>tenfold</u> change in the acidity or basicity. (Another way to do the problem: subtract the pH values <u>first</u>. $10^{-(4.3 - 5.6)}$ = $10^{1.3}$ = <u>20</u>.)

---

1) U. S. News & World Report, July 25, 1988, page 44.

CALCULATING pH FOR SOLUTIONS OF WEAK ACIDS:

One way in which pH is typically used is to describe the acidity of solutions of weak acids. A weak acid, by definition, is not completely dissociated into ions when placed into water. Thus, for weak acids, [acid] $\neq$ [H_3O^+]. Problems such as this must therefore be solved using techniques for dealing with equilibria. The equilibrium constant for the reaction in which a weak acid dissociates into ions has a particular symbol -- K_a, the acid dissociation constant.

Problem: Calculate the pH of a 0.100 M solution of acetic acid (the acid which is a major component of vinegar.). $K_a = 1.76 \times 10^{-5}$ for acetic acid.[1]

Solution: The equilibrium for this reaction is given below:

$$HC_2H_3O_2 \text{ (aq)} + H_2O \text{ (1)} \rightleftharpoons H_3O^+ \text{ (aq)} + C_2H_3O_2^- \text{ (aq)}$$

This is a heterogeneous equilibrium, so no [H_2O] term appears in the equilibrium expression. The equilibrium expression is: $K_a = [H_3O^+][C_2H_3O_2^-]/[HC_2H_3O_2]$. Since we don't know how much of the acetic acid ($HC_2H_3O_2$) dissociates, call that amount "X". If "X" moles of acetic acid dissociates, "X" moles of H_3O^+ and "X" moles of $C_2H_3O_2^-$ will be formed. This information is summarized in the following table:

	Initial Value	Change During Reaction	Final Value
[$HC_2H_3O_2$]	0.100 M	-X	(0.100 - X) M
[$C_2H_3O_2^-$]	0.000 M	+X	X M
[H_3O^+]	0.000 M	+X	X M

Weak acids have small K_a values. This tells us that "X" will be a small number, negligible with respect to 0.100. Hence, using the information above, we get:

$$K_a = [H_3O^+][C_2H_3O_2^-]/[HC_2H_3O_2] = 1.76 \times 10^{-5} = (X)(X)/0.100 = X^2/0.100.$$

$$X = [(1.76 \times 10^{-5})(0.100)]^{1/2} = (1.76 \times 10^{-6})^{1/2} = 1.33 \times 10^{-3} = [H_3O^+].$$

$$pH = -\log[H_3O^+] = -\log(1.33 \times 10^{-3}) = \underline{2.877}.$$

---

1) CRC Handbook of Chemistry and Physics, 56th Edition, 1975-1976.

CALCULATING pH FOR SOLUTIONS OF WEAK BASES:

One way in which pH is typically used is to describe the basicity of solutions of weak bases. A weak base, by definition, does not completely react to form OH^- ions when placed in water. Thus, for weak bases, [base] $\neq$ [OH^-]. In this case, the problem must be solved using techniques for dealing with equilibria. The equilibrium constant for the reaction in which a weak base reacts to form OH^- ions has a particular symbol -- K_b, the base dissociation constant.

Problem: Calculate the pH of a 0.100 M solution of ammonia (yes, the same substance used as a household cleaner!). $K_b = 1.79 \times 10^{-5}$ for ammonia.[1]

Solution: The equilibrium for this reaction is given below:

$$NH_3 \text{ (aq)} + H_2O \text{ (1)} \rightleftharpoons NH_4^+ \text{ (aq)} + OH^- \text{ (aq)}$$

This is a heterogeneous equilibrium, so no [H_2O] term appears in the equilibrium expression. The equilibrium expression is: $K_b = [NH_4^+][OH^-]/[NH_3]$. Since we don't know how much of the ammonia (NH_3) reacts, call that amount "X". If "X" moles of ammonia reacts, "X" moles of NH_4^+ and "X" moles of OH^- will be formed.

Summarizing:

	Initial Value	Change During Reaction	Final Value
[NH_3]	0.100 M	-X	(0.100 - X) M
[NH_4^+]	0.000 M	+X	X M
[OH^-]	0.000 M	+X	X M

The small value of K_b (typical for a weak base) tells us that "X" is a small number and is negligible compared to 0.100. Hence, using the information above, we get:

$K_b = [NH_4^+][OH^-]/[NH_3] = 1.79 \times 10^{-5} = (X)(X)/0.100 = X^2/0.100$.

$X = [(1.79 \times 10^{-5})(0.100)]^{1/2} = (1.79 \times 10^{-6})^{1/2} = 1.34 \times 10^{-3} = [OH^-]$.

$pOH = -\log[OH^-] = -\log(1.34 \times 10^{-3}) = 2.874$.

$pH = 14.000 - pOH = 14.000 - 2.874 = \underline{11.126}$.

---

1) CRC Handbook of Chemistry and Physics, 56th Edition, 1975-1976.

CONJUGATE ACIDS AND CONJUGATE BASES:

Consider the equilibrium for the reaction of a "generic" weak acid, HA, with water: $HA_{(aq)} + H_2O_{(l)} \rightleftharpoons H_3O^+_{(aq)} + A^-_{(aq)}$. If the reaction was to proceed in the <u>reverse</u> direction, the effect of the A^- ion would be to accept a proton (H^+) from the H_3O^+ ion -- that is, the A^- ion would be effectively functioning as a <u>base</u>. For this reason, A^- is referred to as the <u>conjugate base</u> of the acid HA. For a specific example, consider the <u>water</u> molecule in the above reaction -- it accepts a proton from HA, and forms H_3O^+ in the process. Therefore, H_2O can be considered the <u>conjugate base</u> of the acid H_3O^+.

Similarly, consider the equilibrium for the reaction of the A^- ion with water: $A^-_{(aq)} + H_2O_{(l)} \rightleftharpoons HA_{(aq)} + OH^-_{(aq)}$. Again, <u>reversing</u> this reaction shows that HA would donate a proton to the OH^- ion, thus functioning as an <u>acid</u>. Therefore, HA is referred to as the <u>conjugate acid</u> of the base A^-. Again, a specific example would be the <u>water</u> molecule above -- it donates a proton to the A^- ion, forming OH^-. Therefore, H_2O is the <u>conjugate acid</u> of the base OH^-. (Notice that the difference between an acid or base and its conjugate is just one proton!)

The equilibrium expressions for the above equilibria are given below:

$$K_a = [H_3O^+][A^-]/[HA] \qquad\qquad K_b = [HA][OH^-]/[A^-]$$

When these two expressions are multiplied together, something interesting happens:

$$K_a K_b = ([H_3O^+][A^-]/[HA]) \times ([HA][OH^-]/[A^-]) = [H_3O^+][OH^-] = K_w.$$

Thus, <u>for a conjugate acid-base pair</u> (that is, an acid and its conjugate base or a base and its conjugate acid), <u>$K_a \times K_b = K_w = 1.00 \times 10^{-14}$</u>. From this equation, it can be seen that if K_a is large, K_b must be small, and if K_b is large, K_a must be small -- that is, the <u>stronger</u> the acid, the <u>weaker</u> its conjugate base, and the <u>stronger</u> the base, the <u>weaker</u> its conjugate acid. Acids and bases tend to react in such a way as to form their <u>weaker</u> conjugates whenever possible.

CALCULATING pH FOR SOLUTIONS OF SALTS OF WEAK ACIDS:

Problem: Acetic acid ($HC_2H_3O_2$) has a K_a value of 1.76×10^{-5}.[1]
Calculate the pH of a 1.00 M solution of sodium acetate ($NaC_2H_3O_2$) in water.

Solution: When sodium acetate dissolves in water, sodium ions (Na^+) and acetate ions ($C_2H_3O_2^-$) are formed. The acetate ions react with water as shown by the equilibrium: $C_2H_3O_2^-$ (aq) $+ H_2O$ (l) $\rightleftharpoons$ $HC_2H_3O_2$ (aq) $+ OH^-$ (aq). Since OH^- ions are formed in this reaction, the solution will be basic, and its pH will be greater than 7.000. The value of K_b for the acetate ion can be calculated from the equation $K_a K_b = K_w$, since $C_2H_3O_2^-$ is the conjugate base of $HC_2H_3O_2$. Hence:

$$K_b = K_w/K_a = (1.00 \times 10^{-14})/(1.76 \times 10^{-5}) = 5.68 \times 10^{-10}.$$

The equilibrium expression for the above equilibrium is given below:

$$K_b = [HC_2H_3O_2][OH^-]/[C_2H_3O_2^-] = 5.68 \times 10^{-10}$$

To find the concentration of OH^-, the usual methods for dealing with equilibria are used. Call the amount of acetate ion that reacts "X". If "X" moles of acetate ion react, "X" moles of acetic acid and "X" moles of hydroxide ion will be formed.

Summarizing:	Initial Value	Change During Reaction	Final Value
$[C_2H_3O_2^-]$	1.00 M	-X	(1.00 - X) M
$[HC_2H_3O_2]$	0.00 M	+X	X M
$[OH^-]$	0.00 M	+X	X M

Inserting the "Final Value" entries into the above equilibrium expression gives the following (the "X" in the denominator is negligible, due to the small value of K_b):

$$K_b = [HC_2H_3O_2][OH^-]/[C_2H_3O_2^-] = 5.68 \times 10^{-10} = (X)(X)/(1.00) = X^2.$$

$$X = (5.68 \times 10^{-10})^{1/2} = 2.38 \times 10^{-5} = [OH^-].$$

$$pOH = -\log[OH^-] = -\log(2.38 \times 10^{-5}) = 4.623.$$

$$pH = 14.000 - pOH = 14.000 - 4.623 = \underline{9.377}. \quad \text{(Greater than 7.000!)}$$

---

1) CRC Handbook of Chemistry and Physics, 56th Edition, 1975-1976.

BUFFER SOLUTIONS:

Living things cannot tolerate drastic changes in pH in their habitats. For example, in a recent study supported by the Canadian government, "researchers dumped sulfuric acid into a small Ontario lake, systematically raising its acid level 30 times to a pH of 5. When they were done, one third of the resident species had been knocked out -- including the shrimp and crayfish eaten by trout. The emaciated fish stopped reproducing. Relates team leader David Schindler: 'We eventually would have lost every species of fish.'"[1]

Fortunately, oceans and lakes are natural buffer solutions. A buffer solution is one which contains both a weak Brønsted acid and its conjugate base. It is therefore able to resist changes in pH by reacting with either acids or bases which are added to the solution. If an acid is added to a buffer solution, the base already present reacts with it; if a base is added to a buffer solution, the acid already present reacts with it.

The pH of a buffer solution depends upon the concentrations of the acid and conjugate base used to prepare it. In order to see this, consider the generic weak acid equilibrium: $HA_{(aq)} + H_2O_{(l)} \rightleftharpoons H_3O^+_{(aq)} + A^-_{(aq)}$. The equilibrium expression for this equilibrium is: $K_a = [H_3O^+][A^-]/[HA]$. Now, if we take the negative logarithm of each side of this equation, we get the following:

$$-\log K_a = -\log([H_3O^+][A^-]/[HA]) = -\log[H_3O^+] - \log([A^-]/[HA])$$

However, since "$-\log[H_3O^+]$" is simply pH, and "$-\log K_a$" is called pK_a, we get:

$$pK_a = pH - \log([A^-]/[HA]) \quad \text{or, as it is usually written:}$$

$$\underline{pH = pK_a + \log([A^-]/[HA])}.$$

The underlined equation is known as the Henderson-Hasselbalch Equation, and gives us an easy way to find the pH of a buffer solution if pK_a, [HA], and [A$^-$] are known.

1) U. S. News & World Report, July 25, 1988, page 44.

HOW BUFFERS WORK:

Problem: Calculate the pH of a buffer solution made by dissolving 1.00 mole of sodium acetate and 0.100 mole of acetic acid in enough water to make 1.00 liter of solution. Acetic acid has a K_a value of 1.76×10^{-5}.[1]

Solution: Since the volume of the solution is 1.00 liter, $[HC_2H_3O_2]$ = 0.100 M and $[C_2H_3O_2^-]$ = 1.00 M. Now, using the Henderson-Hasselbalch Equation:

$$pH = pK_a + \log([A^-]/[HA]) = -\log K_a + \log([C_2H_3O_2^-]/[HC_2H_3O_2]).$$

$$pH = -\log(1.76 \times 10^{-5}) + \log(1.00\ M/0.100\ M) = 4.754 + 1.000 = \underline{5.754}.$$

Problem: Calculate the pH of the buffer solution formed by adding 0.100 mole of HCl to the above buffer solution.

Solution: HCl is a strong acid, so the effect of adding HCl to the buffer solution is to increase the concentration of H_3O^+. This shifts the equilibrium $HC_2H_3O_2$ (aq) $+ H_2O$ (l) $\rightleftharpoons$ H_3O^+ (aq) $+ C_2H_3O_2^-$ (aq) to the left, increasing $[HC_2H_3O_2]$ and decreasing $[C_2H_3O_2^-]$. If we assume that all of the HCl reacted with acetate ion to form acetic acid, $[HC_2H_3O_2]$ should increase by 0.100 M, and $[C_2H_3O_2^-]$ should decrease by 0.100 M. Therefore, at equilibrium:

$$[HC_2H_3O_2] = 0.100\ M + 0.100\ M = \underline{0.200\ M}.$$

$$[C_2H_3O_2^-] = 1.00\ M - 0.100\ M = \underline{0.90\ M}.$$

Now, simply apply the Henderson-Hasselbalch Equation, as in the previous problem:

$$pH = pK_a + \log([A^-]/[HA]) = -\log K_a + \log([C_2H_3O_2^-]/[HC_2H_3O_2])$$

$$pH = -\log(1.76 \times 10^{-5}) + \log(0.90\ M/0.200\ M) = 4.754 + 0.65 = \underline{5.41}.$$

Notice that the addition of the HCl only changes the pH of the buffer solution from 5.754 to 5.41. This corresponds to about a twofold increase in acidity. If the HCl had been added to neutral (unbuffered) water, the pH would have changed from 7.000 to 1.000 -- a millionfold increase in acidity! Buffers resist changes in pH.

---

1) CRC Handbook of Chemistry and Physics, 56th Edition, 1975-1976.

POLYPROTIC ACIDS AND THEIR SALTS:

A <u>polyprotic acid</u> is an acid whose molecules are each capable of donating more than one proton. Examples of polyprotic acids include sulfuric acid (H_2SO_4), carbonic acid (H_2CO_3), and phosphoric acid (H_3PO_4). Since more than one proton can be donated by each of these acids, more than one equilibrium expression is needed to describe the reactions that take place. For example, the equilibria below describe the reactions of carbonic acid (a <u>diprotic</u> acid) with water:

$$H_2CO_3 \text{ (aq)} + H_2O \text{ (1)} \rightleftharpoons H_3O^+ \text{ (aq)} + HCO_3^- \text{ (aq)} \qquad K_{a1} = 4.30 \times 10^{-7}$$

$$HCO_3^- \text{ (aq)} + H_2O \text{ (1)} \rightleftharpoons H_3O^+ \text{ (aq)} + CO_3^{2-} \text{ (aq)} \qquad K_{a2} = 5.61 \times 10^{-11}$$

The following equilibria describe the reactions of H_3PO_4 (a <u>triprotic</u> acid):

$$H_3PO_4 \text{ (aq)} + H_2O \text{ (1)} \rightleftharpoons H_3O^+ \text{ (aq)} + H_2PO_4^- \text{ (aq)} \qquad K_{a1} = 7.52 \times 10^{-3}$$

$$H_2PO_4^- \text{ (aq)} + H_2O \text{ (1)} \rightleftharpoons H_3O^+ \text{ (aq)} + HPO_4^{2-} \text{ (aq)} \qquad K_{a2} = 6.23 \times 10^{-8}$$

$$HPO_4^{2-} \text{ (aq)} + H_2O \text{ (1)} \rightleftharpoons H_3O^+ \text{ (aq)} + PO_4^{3-} \text{ (aq)} \qquad K_{a3} = 2.2 \times 10^{-13}$$

The K_a values[1] shown are typical for polyprotic acids. From the K_a values, it can be seen that the <u>first</u> proton is donated more readily than the other protons are. (K_{a1} is larger than K_{a2}, which in turn is larger than K_{a3}.) This is due to the fact that the first proton donated leaves a <u>neutral</u> molecule, whereas the second proton donated leaves a <u>negative</u> ion. Protons, being positively charged, are less likely to leave negative ions than neutral molecules. The large differences between the successive K_a values imply that <u>all</u> of the molecules donate their <u>first</u> protons before <u>any</u> of them donate their second or third protons.

The acid salts of polyprotic acids are excellent <u>buffers</u>, since they can either donate protons or accept protons. (Phosphate and carbonate buffers keep our blood at a relatively constant pH of about 7.4.) Buffers like this can be made simply by titrating a polyprotic acid with NaOH until the desired pH is obtained.

---

1) CRC Handbook of Chemistry and Physics, 56th Edition, 1975-1976.

HYDROLYSIS OF SALTS:

When table salt (sodium chloride, NaCl) is dissolved in water, the pH of the resulting solution is neutral -- 7.000. However, other salts form solutions which are not neutral. For example, sodium acetate ($NaC_2H_3O_2$) gives a basic solution when dissolved in water, and ammonium chloride (NH_4Cl) produces an acidic solution. This comes about due to the reaction of water molecules with one of the ions of the salt, producing OH^- ions or H_3O^+ ions. This process is called hydrolysis, which comes from "hydro" (meaning water) and "lysis" (meaning cleavage).

The tendency of some ions to cause hydrolysis to occur is predictable. Ions which are the conjugate bases of weak acids tend to produce basic solutions. Examples include the phosphate ion (PO_4^{3-}), the carbonate ion (CO_3^{2-}), and the acetate ion ($C_2H_3O_2^-$). Ions which are the conjugate acids of weak bases tend to produce acidic solutions. An example is the ammonium ion (NH_4^+). It is noteworthy that two of the above ions (phosphate and carbonate) tend to form ionic compounds which are insoluble in water, but soluble in acids. The basicity of these ions is the property that causes them both to produce hydrolysis and to dissolve in acids. Ions whose ionic compounds are generally water-soluble (such as Na^+, Cl^-, K^+, and NO_3^-) do not cause hydrolysis to occur. They are sometimes called "spectator ions", since they do not react with water. It is noteworthy that the "spectator ions" Cl^- and NO_3^- are the conjugate bases of strong acids. Since strong acids have rather weak conjugate bases, these ions tend not to react with water very strongly.

Problem: Predict the acidity or basicity of: (a) NH_4NO_3, (b) K_2CO_3.

Solution: (a) is acidic, (b) is basic. One way to approach this problem (other than by looking at the information above) is to notice that NH_4NO_3 is formed by the reaction of HNO_3 (a strong acid) and NH_3 (a weak base). Therefore, the solution will be acidic. KOH (strong base) + H_2CO_3 (weak acid) $\longrightarrow$ K_2CO_3.

pH TITRATIONS:

In a typical acid-base titration, an indicator is used to signal the end point. However, the use of an acid-base indicator is not the best way to find the <u>equivalence point</u> -- the point at which <u>all</u> of the acid or base being titrated has reacted. Acid-base indicators are organic substances -- usually, weak Brønsted acids. These indicators have the unique property of changing color upon donation of their protons. However, this color change does not become visible until just <u>after</u> the equivalence point has been reached -- the acid-base reaction under study occurs <u>first</u>, and only <u>then</u> does the indicator react. In addition, it is difficult to determine the exact color of the indicator at the end point just by looking at it -- the eye is not a very good spectroscope!

A more precise method of determining the equivalence point in an acid-base titration is to use a <u>pH meter</u> -- an electronic instrument which measures the pH of a solution accurately. As an example, consider the titration of acetic acid (a weak acid) with sodium hydroxide solution (a strong base). Just before the equivalence point is reached, a small amount of acetic acid will remain unreacted, so the solution will be acidic. Just after the equivalence point, a small amount of sodium hydroxide will be present in the solution, so the solution will be basic. The transition from acidic to basic is marked by a rapid increase in the pH of the solution, which can easily be detected by a pH meter.

It should be noted that just because the acid and the base "neutralize" each other during an acid-base titration, "neutral" solutions (pH = 7.000) are not always obtained at the equivalence point. For example, in the above case, at the equivalence point (where equal amounts (in <u>moles</u>) of sodium hydroxide and acetic acid have been combined), a solution of sodium acetate is obtained. Since acetate ions <u>hydrolyze</u> water molecules, the pH at the equivalence point will be <u>basic</u>.

OXIDATION-REDUCTION REACTIONS:

The rusting of iron and the burning of coal are two common examples of oxidation-reduction reactions (or, for short, "redox" reactions). It is not too surprising that the reaction of something with oxygen is an "oxidation" reaction, but what is "reduction"? To find out, let's look at the above reactions:

$$2 \text{ :Fe:} + 3 \text{ :O:} \longrightarrow 2 \text{ :Fe}^{3+} + 3 \text{ :O:}^{2-} \longrightarrow Fe_2O_3.$$

$$\text{:C·} + 2 \text{ :O:} \longrightarrow \text{:O::C::O:} = CO_2.$$

In the reaction between iron atoms and oxygen atoms, electrons are transferred from the iron atoms to the oxygen atoms. This transfer of electrons from one atom to another is oxidation and reduction in the purest sense. The loss of electrons by an atom is called oxidation; when an atom gains electrons, the process is called reduction. Thus, in the first reaction above, iron atoms are oxidized, and oxygen atoms are reduced. The two processes of oxidation and reduction always take place simultaneously -- if something is oxidized, something else must be reduced. Hence, for an oxidation to take place, something must do the oxidizing. The atom which accepts electrons from another atom, thereby oxidizing it, is called the oxidizing agent. Not surprisingly, oxygen is the oxidizing agent in the reactions above. Similarly, for a reduction to occur, something must do the reducing. The atom that donates electrons to another atom, thereby reducing it, is called the reducing agent. Iron is the reducing agent in the first reaction above.

In the second reaction above, electrons are shared by the carbon and oxygen atoms. However, since oxygen atoms are more electronegative than carbon atoms, we can think of the shared electrons as more the "property" of the oxygen atoms than the carbon atom. Hence, electrons are "transferred", and this is an oxidation-reduction reaction. Notice that the reducing agent (here, carbon) is, itself, oxidized, while the oxidizing agent (here, oxygen) is, itself, reduced.

OXIDATION NUMBERS:

An easy way to keep track of the transferring of electrons (or electron density, for covalent molecules) that occurs during oxidation-reduction reactions is to assign oxidation numbers to each atom present. An atom's __oxidation number__ (or __oxidation state__) is simply the electrical __charge__ that the atom would have if the compound in which that atom is located were completely ionic.

For __ionic__ compounds, assigning oxidation numbers is easy -- just use the charges on the ions themselves! For example, the ionic compound Fe_2O_3 is made up of Fe^{3+} ions and O^{2-} ions. Therefore, in Fe_2O_3, the oxidation number of iron is __+3__, and oxygen is in the __-2__ oxidation state. For covalent compounds, assigning oxidation numbers is a little bit more challenging, since no ions are present. However, the process can be simplified by simply considering all electrons present in covalent bonds to be the "property" of the __more electronegative atom__ in the bond. In other words, assign oxidation numbers to atoms in covalent molecules __as if the compound were completely ionic__. For example, consider the covalent molecule CO_2, whose Lewis Structure is $:\overset{..}{O}::C::\overset{..}{O}:$ (Lone pairs of electrons are the "property" of the atom on which they are located.). If all four electrons in each double bond are assigned to the oxygen atoms, then each oxygen atom will "own" __eight__ electrons. This is the electron configuration for the O^{2-} ion, so oxygen's oxidation number is __-2__. The carbon atom will have lost all four of its valence electrons, so it will be in the __+4__ oxidation state. A good way to check to see if you have assigned the oxidation numbers correctly is to see if the __sum__ of the oxidation numbers is equal to the __total charge__ on the formula as written. For Fe_2O_3, a __neutral__ formula, the sum should equal __zero__, and it does: $(+3) + (+3) + (-2) + (-2) + (-2) = \underline{0}$. For CO_2, another neutral molecule, the sum is: $(+4) + (-2) + (-2) = \underline{0}$. For the polyatomic ion MnO_4^-, Mn is in the +7 oxidation state: $(+7) + (-2) + (-2) + (-2) + (-2) = \underline{-1}$.

BALANCING REDOX EQUATIONS:

In writing balanced equations for oxidation-reduction reactions, _three_ factors must be considered. Obviously, the number of atoms of each element must be equal on both sides of the equation -- that's true for _any_ balanced equation! However, since _ions_ are often involved, the total electrical _charge_ must also be equal on both sides of the equation. Finally, the _total number of electrons lost_ by the reducing agent must _equal_ the _total number of electrons gained_ by the oxidizing agent -- no "free" electrons remain at the end of the reaction.

Problem: Balance the following redox equation, assuming that it occurs under _basic_ conditions: $MnO_4^- + C_2O_4^{2-} \longrightarrow CO_2 + MnO_2$.

Solution: To balance the number of electrons transferred, assign _oxidation numbers_ to each atom present. Oxygen's oxidation number is always _-2_, but the oxidation number of manganese changes from _+7_ (in MnO_4^-) to _+4_ (in MnO_2) during the course of the reaction. Similarly, the oxidation number of each carbon atom changes from _+3_ (in $C_2O_4^{2-}$) to _+4_ (in CO_2) during the course of the reaction. Hence, each manganese atom _gains three electrons_, while each carbon atom _loses one electron_. Therefore, _three_ carbon atoms must be present for each manganese atom present. Adjusting the coefficients to reflect this situation, we get (for now):

$$2 MnO_4^- + 3 C_2O_4^{2-} \longrightarrow 6 CO_2 + 2 MnO_2.$$

To balance the total electrical charge, H^+ ions or OH^- ions must be added. Since we are working under _basic_ conditions here, OH^- ions will be used. The total charge on the left side of the reaction is _-8_ (two -1 ions and three -2 ions = -8), while the total charge on the right side of the equation is _zero_. Adding eight OH^- ions to the right side balances the charge. To balance the atoms of each element, simply add four water molecules (H_2O) to the left side, giving a balanced equation:

$$2 MnO_4^- + 3 C_2O_4^{2-} + 4 H_2O \longrightarrow 6 CO_2 + 2 MnO_2 + 8 OH^-.$$

AN ALTERNATIVE METHOD FOR BALANCING REDOX EQUATIONS:

Another way to balance equations for oxidation-reduction reactions is to separate the overall reaction into two "half-reactions" -- the oxidation half-reaction and the reduction half-reaction. Balancing individual half-reactions is relatively easy, since electrons may be used to balance electrical charges. When each half-reaction has been balanced, the half-reactions are simply added together to obtain the balanced equation. However, before adding the half-reactions, the total numbers of electrons in each half-reaction must be adjusted so that the total number of electrons lost equals the total number of electrons gained.

Problem: Balance the following redox equation, assuming that it occurs under acidic conditions: $NH_3 + Cr_2O_7^{2-} \longrightarrow N_2 + Cr_2O_3$.

Solution: First, separate the overall reaction into two half-reactions:

$$NH_3 \longrightarrow N_2 \qquad\qquad Cr_2O_7^{2-} \longrightarrow Cr_2O_3$$

Next, balance each half-reaction individually, using H_2O to balance the number of oxygen atoms, H^+ to balance the number of hydrogen atoms (H^+ is used because the reaction is occurring under acidic conditions), and electrons (written as e^-) to balance the total electrical charge. Using these steps, and noting that two NH_3 molecules will be needed since there are two nitrogen atoms in N_2, we obtain the following new half-reactions:

$$2\ NH_3 \longrightarrow N_2 + 6\ H^+ + 6\ e^- \qquad \text{(oxidation)}$$
$$6\ e^- + 8\ H^+ + Cr_2O_7^{2-} \longrightarrow Cr_2O_3 + 4\ H_2O \quad \text{(reduction)}$$

Since six electrons are transferred in each half-reaction, the coefficients do not need to be adjusted. Adding these half-reactions together and canceling similar terms (such as $6\ e^-$) on both sides of the arrow, we obtain the balanced equation:

$$2\ H^+ + 2\ NH_3 + Cr_2O_7^{2-} \longrightarrow N_2 + Cr_2O_3 + 4\ H_2O.$$

Notice that this method avoids the use of oxidation numbers.

ELECTROCHEMISTRY:

Sodium is a highly reactive metal. Chlorine is a poisonous, greenish-yellow gas. When sodium and chlorine react, the relatively stable compound sodium chloride (table salt) is formed. The reaction is: $2 Na + Cl_2 \longrightarrow 2 NaCl$. This is an oxidation-reduction reaction -- sodium is oxidized, chlorine is reduced. The reaction proceeds in the direction shown -- the reverse reaction does not occur spontaneously. (In fact, metallic sodium does not occur in nature. Sodium is so reactive that all naturally-occurring sodium exists in the form of sodium salts.) However, it is possible to make the reverse of the above reaction occur by taking electrons away from the Cl^- ions and bringing them into contact with the Na^+ ions. This is done by using a current of electricity as the source of electrons. In a typical Downs cell (used for the industrial production of sodium metal and chlorine gas), sodium chloride is heated until it melts. Two electrodes are then immersed into the molten sodium chloride. The electrodes are connected to a source of electrical current, and a current of electricity is allowed to flow through the electrodes and the molten sodium chloride. (For the current to flow, the ions must be free to move around, which explains why the sodium chloride is melted.) The Na^+ ions migrate toward the negative electrode (called the <u>cathode</u>, because <u>cat</u>ions are attracted to it) and are reduced by the electrons produced by the electric current. The Cl^- ions migrate toward the positive electrode (called the <u>anode</u>, since <u>an</u>ions are attracted to it) and donate electrons to it, thereby becoming oxidized. The reactions which occur during the <u>electrolysis</u> of sodium chloride are shown below:

Cathode: $Na^+ + e^- \longrightarrow Na$ (reduction)

Anode: $2 Cl^- \longrightarrow 2 e^- + Cl_2$ (oxidation)

This process illustrates one aspect of <u>electrochemistry</u>, which is the study of the relationships between electrical energy and chemical (usually <u>redox</u>) reactions.

QUANTITATIVE ELECTROLYSIS:

Electrolysis (from "electro-", meaning electricity, and "-lysis", which means cleavage or breaking apart) is the process by which a compound is converted into the elements from which it is made by using an electrical current to perform oxidation-reduction reactions. One of the most practical uses of electrolysis is in the production of metals from their ions. In order to find practical conditions for doing this, some of the quantitative aspects of electrolysis must be considered.

The unit of electrical current is the ampere. One ampere (or amp, for short) is the electrical current caused by having one coulomb of electrical charge pass through a wire in one second. In equation form, this can be written as either 1 amp = 1 coul/sec or 1 coulomb = (1 amp)(1 sec). Since the particles which carry electrical charges while a current flows are electrons, and since the electrical charge on an individual electron is known to be 1.60×10^{-19} coulombs, we can write the following expression for the charge on one mole (6.02×10^{23}) of electrons: $(6.02 \times 10^{23}$ electrons/mole$)(1.60 \times 10^{-19}$ coul/electron$)$ = 96,500 coulombs/mole e^-. This number is sometimes called Faraday's constant (or one Faraday, symbol $\underline{F}$) in honor of the English physicist Michael Faraday, who discovered electromagnetics.

Problem: What mass of copper metal is produced by passing a current of 1.00 amp for 5.00 minutes through a solution of Cu^{2+} ions?

Solution: Since 1 coulomb = (1 amp)(1 sec), the electrical charge is:

(1.00 amp)(5.00 min)(60 sec/min) = 300 (amp x sec) = 300 coulombs.

Now, since 96,500 coulombs = 1 mole of electrons, the amount of electrons is:

(300 coulombs)(1 mole of electrons/96,500 coulombs) = 3.11×10^{-3} moles.

The "half-reaction" which occurs is: $Cu^{2+}_{(aq)} + 2\ e^- \longrightarrow Cu_{(s)}$. Hence:

$(3.11 \times 10^{-3}$ moles $e^-)$(1 mole Cu/2 moles e^-) = 1.55×10^{-3} moles Cu.

$(1.55 \times 10^{-3}$ moles Cu$)$(63.54 g/mole Cu) = 0.0988 grams of Cu.

APPLICATIONS OF ELECTROLYSIS:

The process of electrolysis is useful for producing many things that we use routinely, or that affect our lives in other ways. One of the most notable of these is the production or purification of various metals by electrolysis. Sodium metal is produced from molten sodium chloride by electrolysis in the <u>Downs</u> cell. (A "cell" is the term used to describe any electrochemical process, or the vessel in which the process occurs.) The Downs process also produces chlorine gas, which can be used in the production of pesticides. The sodium chloride used in the Downs process comes from an abundant source: evaporation of sea water! Sea water also contains magnesium ions, and magnesium metal can be produced by electrolysis of $MgCl_2$ in a process called the <u>Dow</u> process. Copper and aluminum metals can also be produced in this way, although the Hall process for producing aluminum is expensive. (It's cheaper to recycle aluminum cans than to mine and produce new aluminum!)

A variation on this basic theme is the process known as <u>electroplating</u>, in which thin metal coatings are applied to a metal surface. This is done simply by setting up an electrolysis cell, using the metal surface to be coated as the cathode and immersing it in a solution which contains ions of the metal to be used as the coating. "Gold-plated" jewelry, "silver" eating utensils, and the "chrome" finish on automobile grillwork ("chrome" is short for <u>chromium</u>) are all produced by this basic process.

Pure water does not conduct electricity, but sea water contains enough ions to allow it to conduct electricity quite well. Electrolysis of sea water produces three major products: chlorine gas, hydrogen gas (used in the manufacture of fertilizers), and sodium hydroxide (used in making soaps). The reactions are:

<u>Anode</u>: $2 \text{ Cl}^-_{(aq)} \longrightarrow \text{Cl}_{2 (g)} + 2 e^-$ (oxidation)

<u>Cathode</u>: $2 \text{ H}_2\text{O}_{(l)} + 2 e^- \longrightarrow \text{H}_{2 (g)} + 2 \text{ OH}^-_{(aq)}$ (reduction)

GALVANIC CELLS:

Different kinds of batteries are used as energy sources in many of the electrically-based instruments we use routinely. Batteries provide energy for our cars, our flashlights, our wristwatches, and our calculators. The energy that a battery supplies comes from chemical reactions that take place inside the battery. Batteries are sometimes called galvanic cells (or voltaic cells). They differ from electrolytic cells in that galvanic cells use a chemical reaction (usually a redox reaction) to supply electrical energy, whereas exactly the opposite is true in an electrolytic cell -- electrical energy is used to make a chemical reaction occur.

In a typical galvanic cell, the electrodes and solutions for each of the two "half-reactions" are kept separate from each other. (This is in contrast to the situation in a typical electrolysis cell, in which the electrodes and the solutions are often in direct contact with each other.) The only connection between the two "half-reactions" is usually a wire which connects the electrodes. Therefore, for the transfer of electrons from one "half-reaction" to the other to occur, the electrons must flow through the wire connecting the electrodes. The flow of electrons through a wire is commonly called an electric current. Thus, an oxidation-reduction reaction can be used to generate a current of electricity.

The force which "pushes" electrons through a wire is called the electromotive force (or e.m.f., for short). This force is measured in units called volts. (1 volt = 1 Joule/1 coulomb.) Therefore, if a voltmeter is connected to each of the two electrodes, the voltage (sometimes called the potential difference) between the two "half-reactions" can be measured. This voltage is sometimes called the cell potential (symbol E_{cell}) of a galvanic cell. The standard cell potential (symbol E^0_{cell}) of a galvanic cell is the cell potential of the galvanic cell when measured at 25 ^{0}C, 1.00 atmosphere, and concentrations of all ions equal to 1.00 M.

STANDARD REDUCTION POTENTIALS:

It is not possible to measure the voltage produced by a single "half-reaction", since oxidation and reduction always happen simultaneously. When faced with a situation like this, scientists do the next best thing -- they declare one "half-reaction" to be the standard against which all other "half-reactions" are measured. The standard hydrogen electrode (SHE), for which the "half-reaction" is $2 H^+$ (aq) $+ 2 e^- \rightleftharpoons H_2$ (g), consists of a platinum electrode surrounded by hydrogen gas and immersed in a solution which contains H^+ ions -- that is, an acid solution. The potential of this "half-reaction" has been arbitrarily declared to be 0.0000 volts.[1] The potential of any other "half-reaction" can thus be measured by connecting its electrode to the SHE through a voltmeter. The reading on the voltmeter is the voltage (or potential difference) between the two electrodes, and is called the standard reduction potential (symbol E^0) for the "half-reaction" in question if the measurement is made at 25 ^{0}C, 1.00 atm, and [ion] = 1.00 M.

Notice that the standard reduction potential is written as a reduction, with the ions on the left gaining electrons. Notice also that it is written as an equilibrium, since in any given galvanic cell, any given "half-reaction" may be either the oxidation or the reduction. To find the standard cell potential of a particular galvanic cell, simply find the difference between the standard reduction potentials of the two "half-reactions" in the galvanic cell. (That's why the voltage is sometimes called the "potential difference"!) In equation form, this is: $E^0_{cell} = E^0_{reduction} - E^0_{oxidation}$. This equation can also be used to tell which "half-reaction" is the oxidation and which "half-reaction" is the reduction, since all functioning galvanic cells have a positive value of E^0_{cell}. If the subtraction above produces a negative E^0_{cell} value, the cathode and anode should be reversed.

---

1) CRC Handbook of Chemistry and Physics, 56th Edition, 1975-1976.

USING STANDARD REDUCTION POTENTIALS:

Problem: Given the following standard reduction potentials[1]:

$Cu^{2+} + 2 e^- \rightleftharpoons Cu$ $E^0 = +0.3402$ V

$Zn^{2+} + 2 e^- \rightleftharpoons Zn$ $E^0 = -0.7628$ V

calculate E^0_{cell} for the following cell: Zn / Zn^{2+} (1.00 M) // Cu^{2+} (1.00 M) / Cu.

Solution: The notation above is sometimes used to describe galvanic cells. The advantage of this notation is that it makes it somewhat easier to see the half-reactions that occur in the cell. The half-reactions can be understood by simply reading from left to right -- Zn is converted into Zn^{2+}, and Cu^{2+} becomes Cu. Thus, the overall cell reaction is: $Zn_{(s)} + Cu^{2+}_{(aq)} \rightarrow Zn^{2+}_{(aq)} + Cu_{(s)}$. Since zinc is being oxidized, zinc is the anode in this cell; since copper is being reduced, copper is the cathode in this cell. To find E^0_{cell}, just subtract the two standard reduction potentials above, following the equation below:

$E^0_{cell} = E^0_{reduction} - E^0_{oxidation} = E^0_{cathode} - E^0_{anode} = E^0_{Cu} - E^0_{Zn}$.

$E^0_{cell} = (+0.3402$ V$) - (-0.7628$ V$) = \underline{+1.103 \text{ volts}}$.

Problem: Given the following standard reduction potentials[1]:

$Fe^{2+} + 2 e^- \rightleftharpoons Fe$ $E^0 = -0.409$ V

$Cr^{3+} + 3 e^- \rightleftharpoons Cr$ $E^0 = -0.74$ V

calculate E^0_{cell} for the following cell: Fe / Fe^{2+} (1.00 M) // Cr^{3+} (1.00 M) / Cr.

Solution: It's tempting to try to make some kind of "correction" for the fact that different numbers of electrons are being transferred in the two half-reactions, but don't do this -- just find the potential difference, as above:

$E^0_{cell} = E^0_{redn.} - E^0_{oxdn.} = E^0_{Cr} - E^0_{Fe} = (-0.74$ V$) - (-0.409$ V$) = \underline{-0.33 \text{ V}}$.

The negative value of E^0_{cell} tells us that the cell notation is incorrect. Actually, chromium is the anode, iron is the cathode, and the real value of E^0_{cell} is +0.33 V.

---

1) CRC Handbook of Chemistry and Physics, 56th Edition, 1975-1976.

CELL POTENTIALS, FREE ENERGIES, AND EQUILIBRIUM CONSTANTS:

The free energy of a reaction, ΔG_r^0, measures the amount of useful work that a chemical reaction can do. The amount of work that a galvanic cell can do is a function of the cell potential, E_{cell}^0. However, since the cell potential depends on the chemical reactions taking place in the cell, these two functions can be equated: $\underline{\Delta G_r^0 = -nFE_{cell}^0}$. Here, F is Faraday's constant, and n is the number of moles of electrons being transferred in the balanced equation of the cell reaction.

The free energy of a reaction is also related to the reaction's equilibrium constant, by the equation: $\Delta G_r^0 = -RT \ln K_{eq}$. Therefore, the right side of this equation is equal to the right side of the equation in the preceding paragraph: $-nFE_{cell}^0 = -RT \ln K_{eq}$. Rearranging this equation allows us to express E_{cell}^0 as a function of other variables: $E_{cell}^0 = \frac{RT}{nF} \ln K_{eq}$. Since the values of R, T, and F are known, and since $\ln X = 2.303 \log X$, this equation can be rewritten as:

$$E_{cell}^0 = \frac{(8.314 \text{ Joules/K mole})(298 \text{ K})(2.303)}{n(96,500 \text{ coulombs/mole})} \log K_{eq} = \frac{0.0592 \text{ V}}{n} \log K_{eq}.$$

Problem: Calculate ΔG_r^0 and K_{eq} for the following equilibrium:

$$Zn_{(s)} + Cu^{2+}_{(aq)} \rightleftharpoons Zn^{2+}_{(aq)} + Cu_{(s)}.$$

The value of E_{cell}^0 is +1.103 V for the cell: $Zn / Zn^{2+} // Cu^{2+} / Cu.$[1]

Solution: Use the above equations. Notice that the units cancel!

$\Delta G_r^0 = -nFE_{cell}^0 = -(2 \text{ moles } e^-)(96,500 \text{ coulombs/mole } e^-)(+1.103 \text{ V})$

$\underline{\Delta G_r^0 = -2.13 \times 10^5 \text{ Joules}} = \underline{-213 \text{ kJ}}$. (1 volt = 1 Joule/1 coulomb.)

In this case, n = 2, since two moles of electrons are transferred in the reaction.

$E_{cell}^0 = \frac{0.0592 \text{ V}}{n} \log K_{eq} = +1.103 \text{ V} = \frac{0.0592 \text{ V}}{2} \log K_{eq}.$

$\log K_{eq} = (2)(+1.103 \text{ V})/(0.0592 \text{ V}) = 37.3.$

$K_{eq} = 10^{37.3} = \underline{2 \times 10^{37}}$. The large K_{eq} value indicates a spontaneous reaction, as do the negative ΔG_r^0 value and the positive value of E_{cell}^0.

---

1) CRC Handbook of Chemistry and Physics, 56th Edition, 1975-1976.

HOW CELL POTENTIALS DEPEND ON ION CONCENTRATIONS:

No battery keeps on providing power forever without recharging. This is due to the fact that the energy provided by a battery comes from the oxidation-reduction reactions taking place inside the battery. Once the reactants are used up, the battery can no longer provide power. Therefore, it can be seen that the amount of energy available from a battery -- that is, the cell potential of a galvanic cell -- depends on the concentrations of the reactants of the oxidation-reduction taking place in the battery. This is shown by the following equation: $E_{cell} = E^o_{cell} - \frac{0.0592 \text{ V}}{n} \log Q$, called the Nernst Equation in honor of the German physical chemist Walther Nernst, who won the 1920 Nobel Prize in Chemistry. In the Nernst Equation, Q is the reaction quotient for the equilibrium reaction that takes place in the galvanic cell -- $[products]^p/[reactants]^r$.

Problem: Given that E^o_{cell} = +1.103 V for the following cell[1]:

Zn / Zn^{2+} (1.00 M) // Cu^{2+} (1.00 x 10^{-4} M) / Cu

find the value of E_{cell}, given the ion concentrations above.

Solution: The equilibrium reaction that takes place in this cell is: $Zn_{(s)} + Cu^{2+}_{(aq)} \rightleftharpoons Zn^{2+}_{(aq)} + Cu_{(s)}$. This is a heterogeneous equilibrium, for which the reaction quotient is: $Q = [Zn^{2+}]/[Cu^{2+}]$. Since two moles of electrons are transferred in the balanced equation, n = 2. Thus, we get:

$E_{cell} = E^o_{cell} - \frac{0.0592 \text{ V}}{n} \log Q = E^o_{cell} - \frac{0.0592 \text{ V}}{n} \log ([Zn^{2+}]/[Cu^{2+}])$.

$E_{cell} = (+1.103 \text{ V}) - (0.0592 \text{ V}/2) \log ((1.00 \text{ M})/(1.00 \times 10^{-4} \text{ M}))$.

$E_{cell} = (+1.103 \text{ V}) - (0.0592 \text{ V}/2)(+4.000) = +1.103 \text{ V} - (+0.118 \text{ V})$.

$E_{cell} = \underline{+0.985 \text{ V}}$ -- a bit less than E^o_{cell}. As the battery discharges, the system comes to equilibrium, at which point Q = K_{eq}. Therefore, at equilibrium, $E_{cell} = E^o_{cell} - \frac{0.0592 \text{ V}}{n} \log K_{eq} = E^o_{cell} - E^o_{cell} = \underline{0}$. (This is a "dead" battery!)

1) CRC Handbook of Chemistry and Physics, 56th Edition, 1975-1976.

APPLICATIONS OF GALVANIC CELLS:

People use many different types of batteries every day. One of the most common types of batteries is the car battery, also known as the lead storage battery. The electrodes in the lead storage battery consist of metallic lead and lead(IV) oxide, each of which is converted into lead sulfate as the battery runs. The cell notation for this battery is: $Pb_{(s)} / PbSO_{4(s)} // PbO_{2(s)} / PbSO_{4(s)}$. The source of the sulfate ions is a dilute solution of sulfuric acid that serves as the electrolyte in this battery. The fluid nature of the electrolyte allows the ions in solution to move freely, which enables this battery to be recharged. In a similar fashion, the nickel-cadmium batteries used in calculators may be recharged, since the electrolyte present is a solution of hydroxide ions. In these batteries, the electrodes are metallic cadmium and and nickel(IV) oxide. The cell notation for this battery is: $Cd_{(s)} / Cd(OH)_{2(s)} // NiO_{2(s)} / Ni(OH)_{2(s)}$.

By contrast, common flashlight batteries cannot be recharged. These batteries are sometimes called zinc-carbon dry cells, and their "dryness" is what prevents them from being recharged. The electrodes in the zinc-carbon dry cell are the zinc casing of the battery and the rod of carbon that extends through the center of the battery. The space between the electrodes is filled with a thick paste that contains manganese(IV) oxide, among other things. The cell notation for this battery is: $Zn / Zn^{2+} // MnO_2 / Mn^{3+}$. The Zn^{2+} and Mn^{3+} ions are held in place by the thick paste, and are not free to move. This hinders any efforts to recharge a flashlight battery, since in order to make the redox reaction run in the reverse direction, an external source of electricity is used to reverse the polarity of the electrodes. The ions must be free to migrate to the opposite electrode from the electrode at which they are formed for the reverse reaction to proceed. This is not the case in the zinc-carbon "dry" cell.

THE CHEMISTRY OF HYDROGEN:

Hydrogen was discovered in 1766 by the English chemist Henry Cavendish. It is by far the most abundant element in the universe -- much of the mass of stars is hydrogen -- but only traces of it occur in the earth's atmosphere. Hydrogen consists of three major isotopes: _protium_ ($_1^1H$), _deuterium_ ($_1^2H$), and _tritium_ ($_1^3H$). Naturally-occurring hydrogen is 99.98% protium, with deuterium making up most of the remaining 0.02%. Tritium, which only occurs naturally in trace amounts, can be made in nuclear reactors. It is radioactive, and is used in making hydrogen bombs.

Hydrogen can be prepared in many ways. The name "hydrogen" comes from the Greek words "hydro" (meaning _water_) and "genes" (meaning _origin_), so it's not too surprising that one way to generate hydrogen is by the electrolysis of water:

$$2\ Cl^-_{(aq)} + 2\ H_2O_{(l)} \longrightarrow H_2_{(g)} + Cl_2_{(g)} + 2\ OH^-_{(aq)}.$$

Hydrogen can also be prepared by the reaction of water with reactive metals such as sodium, or by the reaction of acids with less reactive metals such as zinc:

$$2\ Na_{(s)} + 2\ H_2O_{(l)} \longrightarrow H_2_{(g)} + 2\ NaOH_{(aq)}.$$

$$Zn_{(s)} + 2\ HCl_{(aq)} \longrightarrow H_2_{(g)} + ZnCl_2_{(aq)}.$$

Industrially, hydrogen is produced by the reaction of superheated steam with _coke_, a residue of the distillation of coal that is essentially pure carbon:

$$H_2O_{(g)} + C_{(s)} \longrightarrow H_2_{(g)} + CO_{(g)}.$$

(The mixture of H_2 and CO formed by this process is sometimes called "water gas".)

Hydrogen has been used in the past in "lighter-than-air" balloons, since its density is less than that of air. However, its use for this purpose has been discontinued since the explosion of the Hindenburg, the German hydrogen-filled dirigible, in 1937. Hydrogen reacts exothermically with oxygen to form water (see below), so the much less reactive helium is now used to make balloons fly.

$$2\ H_2_{(g)} + O_2_{(g)} \longrightarrow 2\ H_2O_{(l)}. \qquad \Delta H^o_r = -68.32 \text{ kcal/mole.}$$

REACTIONS OF HYDROGEN:

Hydrogen reacts with the more reactive metals to form metal <u>hydrides</u> --
that is, compounds in which the oxidation state of hydrogen is <u>-1</u>. Examples:

$$2 \; Na \; + \; H_2 \; \longrightarrow \; 2 \; NaH. \qquad Ca \; + \; H_2 \; \longrightarrow \; CaH_2.$$

Metal hydrides are strong enough <u>bases</u> to react with water. For this reason, they
are sometimes used as drying agents. Hydrogen gas is a product of this reaction:

$$NaH \; + \; H_2O \; \longrightarrow \; H_2 \; + \; NaOH.$$

Notice that the relatively "weak" base NaOH is formed. Thus, NaH is a <u>strong</u> base.

Hydrogen reacts with nonmetals to form compounds in which the oxidation
state of hydrogen is <u>+1</u>. If these compounds react with water, they react as <u>acids</u>:

$$H_2 \; + \; Cl_2 \; \longrightarrow \; 2 \; HCl. \qquad HCl \; + \; H_2O \; \longrightarrow \; H_3O^+ \; + \; Cl^-.$$

$$2 \; H_2 \; + \; C \; \longrightarrow \; CH_4. \qquad CH_4 \; + \; H_2O \; \longrightarrow \; \text{no reaction.}$$

HCl is a <u>strong</u> acid, but CH_4 is so weak that it's not considered to be an acid.

The above reaction between hydrogen and carbon to produce methane (CH_4)
is the reaction that occurs during the industrial process known as the <u>gasification</u>
of coal. Methane is the major component of natural gas, which burns somewhat more
cleanly than coal does. Therefore, this process will probably be important in the
future as the problem of air pollution is seriously addressed. Methane <u>can</u> react
with water, if the water is in the form of superheated steam:

$$CH_4 \; (g) \; + \; H_2O \; (g) \; \longrightarrow \; 3 \; H_2 \; (g) \; + \; CO \; (g).$$

Mixtures of H_2 and CO such as the one formed by this process are sometimes called
"water gas". They are important industrially in the production of <u>methanol</u> (CH_3OH),
which is another clean-burning fuel. This is known as the <u>Fischer-Tropsch Process</u>:

$$CO \; (g) \; + \; 2 \; H_2 \; (g) \; \longrightarrow \; CH_3OH \; (1).$$

Other major uses of hydrogen include the production of ammonia for use
in fertilizers and the conversion of vegetable oils into vegetable shortenings.

THE CHEMISTRY OF CARBON:

Elemental carbon occurs in nature in three basic forms: amorphous carbon (such as coal, charcoal, and lampblack), graphite (the "lead" in pencils), and diamond (gemstones). The word "amorphous" means "without form or shape", but the other two forms have very definite crystal structures. Each carbon atom in graphite is sp^2-hybridized, with the result that graphite tends to form layers of carbon atoms. The layers have a tendency to slide past each other, which gives graphite its soft, "slippery" feel. Each carbon atom in diamond is sp^3-hybridized, with the result that the carbon atoms in diamond are tightly clustered and held firmly in place, making diamond among the hardest substances known.

The three major isotopes of carbon are carbon-12 ($^{12}_{6}C$), carbon-13 ($^{13}_{6}C$), and carbon-14 ($^{14}_{6}C$). Naturally-occurring carbon is 98.89% carbon-12, with most of the other 1.11% being carbon-13. Carbon-14, which only occurs naturally in trace amounts, is radioactive. The relative amounts of carbon-14 present in ancient artifacts can be used to estimate the age of the artifact. For example, this was done with the Shroud of Turin, establishing it as having been fabricated during the Middle Ages, and therefore certifying that it was not the burial shroud of Christ.

Much of what is usually thought of as the chemistry of carbon is called organic chemistry. Carbon and carbon compounds make up less than 0.03% of the earth's crust, but make up about 17.5% of the human body. The word "organic" comes from "organism" -- organic chemistry is the chemistry of the compounds that make up living organisms. The reason that carbon is so important to living things is that carbon atoms have a strong tendency to form covalent bonds to each other, with the resulting molecules usually consisting of long strands of carbon atoms (along with other atoms, such as hydrogen, oxygen, and nitrogen). Long, flexible molecules like these are ideal for making many of the tissues and membranes found in living things.

OXIDES OF CARBON:

The two most common oxides of carbon are <u>carbon monoxide</u> (CO) and <u>carbon dioxide</u> (CO_2). Carbon monoxide is one component of automobile exhaust, and can be prepared industrially by the reaction of superheated steam with hot carbon. It is used in the industrial synthesis of <u>methanol</u> (CH_3OH), the alcohol component in the fuel mixture known as "gasohol". Carbon monoxide is a poisonous gas. This effect comes from the fact that the brain needs oxygen to survive, and oxygen is transported from the lungs to the brain by binding to iron atoms in the hemoglobin in the bloodstream. Carbon monoxide also binds to the iron in hemoglobin, but the iron-carbon bond doesn't break when the hemoglobin gets to the brain (unlike the iron-oxygen bond, which <u>does</u> break to allow the O_2 molecule to be "dropped off" at the brain). Thus, carbon monoxide prevents the hemoglobin from transporting oxygen to the brain, causing the victim to die of asphyxiation.

Carbon dioxide occurs naturally in the earth's atmosphere. We exhale carbon dioxide, which is taken in by plants and converted into simple sugars (such as <u>glucose</u>, $C_6H_{12}O_6$) by the process known as <u>photosynthesis</u>. The reaction is:

$$6\ CO_2\ +\ 6\ H_2O\ +\ light\ \longrightarrow\ C_6H_{12}O_6\ +\ 6\ O_2.$$

Carbon dioxide in the atmosphere tends to absorb infrared radiation (given off by the earth) and radiate it back toward the earth in the form of heat. This is known as the "greenhouse effect". In the late 1980's, the problem of global warming due to increasing amounts of CO_2 in the atmosphere (from population growth, increasing consumption of carbon-based fuels such as coal, oil, gasoline, and natural gas, and destruction of rain forests) was a matter of increasing public concern.[1] Frozen CO_2 is known as "dry ice", since it sublimes. CO_2 reacts with water to form H_2CO_3, which is the source of carbonate (CO_3^{2-}) ions in chalk, coral, limestone, and marble.

---

1) TIME Magazine, July 4, 1988, page 18.

OTHER CARBON COMPOUNDS:

Carbides are compounds in which carbon has a negative oxidation number. For example, the oxidation number of carbon is -1 in calcium carbide (CaC_2), which is formed by the reaction of calcium oxide (lime) and carbon at high temperatures:

$$CaO_{(s)} + 3 C_{(s)} + heat \longrightarrow CaC_{2 (s)} + CO_{(g)}.$$

Calcium carbide reacts with water to form acetylene (C_2H_2), which is used as a fuel in torches and lanterns. The reaction is: $CaC_2 + 2 H_2O \longrightarrow Ca(OH)_2 + C_2H_2$.

Carbon compounds which contain chlorine are generally toxic and possibly carcinogenic (cancer-causing). Chloroform ($CHCl_3$) was a common anaesthetic, and carbon tetrachloride (CCl_4) was routinely used as a cleaning fluid, but both have been avoided since their possible harmful effects became known. Phosgene ($COCl_2$), prepared by the reaction of carbon monoxide with elemental chlorine, was used in World War One as a poisonous, lachrymating (tear-producing) gas. Compounds known collectively as chlorofluorocarbons (CFC's for short) are the "freons" used in air conditioners, refrigerators, and aerosol spray cans as refrigerants and propellants. Their harmful effects on the earth's ozone layer has become a topic of some concern.

The most notable carbon compound which contains sulfur is carbon disulfide (CS_2), which is used in the production of the synthetic material Rayon. Cellulose is obtained by breaking down wood fibers using dilute sulfuric acid, and the addition of carbon disulfide produces a material known as cellulose xanthate. Rayon is simply cellulose xanthate that has been purified and spun into fibers.

The most familiar carbon compound which contains nitrogen is the toxic gas hydrogen cyanide (HCN). Cyanide ions ($:C \equiv N:$) have similar Lewis Structures to molecules of carbon monoxide ($:C \equiv O:$), and thus act as poisons in a similar fashion. Organic molecules which contain cyanide are called nitriles. Acrylonitrile is used to make Orlon; adiponitrile is the raw material used to make Nylon.

70

THE CHEMISTRY OF SILICON:

Silicon was discovered by the Swedish chemist Jöns Jakob Berzelius in 1824. It is the second most abundant element on earth, making up over 25% of the earth's crust. Silicon has three major isotopes: silicon-28 ($^{28}_{14}Si$), silicon-29 ($^{29}_{14}Si$), and silicon-30 ($^{30}_{14}Si$). Silicon-28 makes up over 90% of naturally-occurring silicon. Elemental silicon does not occur in nature, but can be made by heating silicon dioxide (SiO_2) in an electric arc furnace with carbon as a reducing agent:

$$SiO_2 \text{ (s)} + 2 \text{ C (s)} + heat \longrightarrow Si \text{ (s)} + 2 \text{ CO (g)}.$$

Silicon dioxide occurs naturally in many forms, the most abundant and least expensive of which is <u>sand</u>. Quartz, flint, amethyst, and opal are other forms of silicon dioxide which occur naturally. <u>Silicates</u> are similar to the above minerals in the sense that they contain silicon in its <u>+4</u> oxidation state. However, silicates contain SiO_4^{4-} ions. An example is zirconium silicate ($ZrSiO_4$), which is known as <u>zircon</u>. Zirconium silicate has also been used in some toothpastes as an abrasive agent. <u>Polysilicates</u> contain ions made from several silicon atoms as well as oxygen atoms. Talc, mica, asbestos, and beryl are examples of polysilicates.

Unlike carbon atoms, silicon atoms do not tend to form covalent bonds to each other. However, long, strand-like molecules similar to those formed by carbon atoms may be formed by alternating silicon atoms with oxygen atoms. These molecules are called <u>silicones</u>. Silicones are used as lubricating oils, caulking agents, and ceramics. ("Silly Putty" is also a silicone -- hence its name!)

Silicon dioxide is used in the manufacture of glass and cement. For example, "soda-lime" glass is prepared by heating a mixture of SiO_2, Na_2CO_3, and $CaCO_3$. "Portland cement" is prepared by heating a mixture of SiO_2, Al_2O_3, and $CaCO_3$. In addition, silicon's semiconductor properties make it useful in the manufacture of transistors -- so much so that "Silicon Valley" is named for it!

THE CHEMISTRY OF NITROGEN:

Nitrogen was first discovered in 1772 by the Scottish chemist Daniel Rutherford. Nitrogen gas makes up over 75% of the earth's atmosphere, from which it is obtained in pure form by cooling it to a liquid and distilling it. Nitrogen has two naturally-occurring isotopes: nitrogen-14 ($^{14}_{7}$N) and nitrogen-15 ($^{15}_{7}$N). Nitrogen-14 makes up over 99.6% of naturally-occurring nitrogen. Liquid nitrogen is extremely cold (the boiling point of nitrogen is approximately -196 ^{O}C) and is used as a refrigerant for the freezing of food products and transportation of foods.

One of the major industrial uses of nitrogen is in the production of ammonia (NH_3). The conversion of nitrogen gas into ammonia is called nitrogen fixation, and is extremely important to living things which need nitrogen to grow. Certain microorganisms can accomplish nitrogen fixation, but the method for doing this on the larger industrial scale was desired in order to increase the amount of nitrogen that could be added to the soil in a short period of time. This method was developed by the German chemist Fritz Haber (who won the 1918 Nobel Prize in Chemistry for this work), and is known as the Haber Process. The reaction is:

$$N_2 \text{ (g)} + 3 \text{ } H_2 \text{ (g)} \rightleftharpoons 2 \text{ } NH_3 \text{ (g)}.$$

High pressures are used in order to shift this equilibrium to the right. Ammonia can be used as a fertilizer directly by liquefying it, or it can be allowed to react with nitric acid (HNO_3) to form ammonium nitrate (NH_4NO_3), a crystalline solid which is easier to handle than liquid ammonia. Nitric acid can be prepared from ammonia by the three-step process (known as the Ostwald Process) shown below:

$$4 \text{ } NH_3 \text{ (g)} + 5 \text{ } O_2 \text{ (g)} \longrightarrow 4 \text{ } NO \text{ (g)} + 6 \text{ } H_2O \text{ (g)}.$$

$$2 \text{ } NO \text{ (g)} + O_2 \text{ (g)} \longrightarrow 2 \text{ } NO_2 \text{ (g)}.$$

$$3 \text{ } NO_2 \text{ (g)} + H_2O \text{ (l)} \longrightarrow 2 \text{ } HNO_3 \text{ (l)} + NO \text{ (g)}.$$

Nitric acid is also used in the production of the explosives nitroglycerin and TNT.

THE MANY OXIDATION STATES OF NITROGEN:

Ammonia (NH_3) contains nitrogen in its <u>-3</u> oxidation state. Nitric acid (HNO_3) contains nitrogen in its <u>+5</u> oxidation state. These two compounds represent the extremes of the many oxidation states which nitrogen occupies in its various compounds. The following is a summary of the other oxidation states of nitrogen, along with some examples of important nitrogen compounds for each oxidation state.

<u>Other -3</u>: Nitrogen is an essential component of <u>amino acids</u> (such as <u>glycine</u>, $H_2N-CH_2-CO_2H$), the "building blocks" from which proteins are made. NI_3 is a contact explosive -- it detonates violently if enough force is applied.

<u>-2</u>: Ammonia reacts with hypochlorite ions to form <u>hydrazine</u> (N_2H_4):

$$NH_{3\ (aq)} + OCl^-_{\ (aq)} \longrightarrow OH^-_{\ (aq)} + NH_2Cl_{\ (aq)}.$$

$$NH_2Cl_{\ (aq)} + NH_{3\ (aq)} \longrightarrow HCl_{\ (aq)} + N_2H_{4\ (aq)}.$$

Derivatives of hydrazine are used as rocket fuels. (Hypochlorite ions are found in household bleach, but don't mix ammonia and bleach at home -- hydrazine is toxic!)

<u>-1</u>: <u>Hydroxylamine</u> (H_2N-OH) reacts with certain organic compounds to form compounds called <u>oximes</u>, which helps scientists identify these compounds.

$$CH_2=O + H_2N-OH \longrightarrow H_2O + CH_2=N-OH \longleftarrow \text{(the oxime)}$$

<u>+1</u>: Nitrous oxide (N_2O) is used as an anaesthetic -- it's commonly known as "laughing gas"! It's also used as the "whipping" agent in "self-whipping" cream -- when you push the nozzle of the whipped cream canister, the nitrous oxide comes out along with the cream, causing the cream to "froth" or "whip".

<u>+3</u>: Nitrous acid (HNO_2) is used in the making of <u>azo dyes</u> -- colorful organic compounds which contain a N=N bond and are used in the textiles industry.

<u>+2, +4</u>: NO (+2) and NO_2 (+4) are air pollutants which can react with moisture in the air to form acids, which fall to earth as <u>acid rain</u>.

<u>Other +5</u>: KNO_3 is a component of gunpowder and fireworks.

THE CHEMISTRY OF PHOSPHORUS:

Phosphorus was discovered in 1669 by the German chemist Hennig Brand. It has only one naturally-occurring isotope, phosphorus-31 ($^{31}_{15}$P). Phosphorus has three major allotropes: <u>white</u> phosphorus, <u>red</u> phosphorus, and <u>black</u> phosphorus. Elemental phosphorus does not occur naturally; the principal sources of phosphorus are various phosphate minerals, such as <u>phosphorite</u> (calcium phosphate, $Ca_3(PO_4)_2$), which is found in the Soviet Union and several U.S. states. White phosphorus (P_4) can be prepared from phosphorite by heating it in the presence of carbon and silica:

$$2 \ Ca_3PO_4 \ + \ 10 \ C \ + \ 6 \ SiO_2 \ + \ heat \ \longrightarrow \ P_4 \ + \ 10 \ CO \ + \ 6 \ CaSiO_3.$$

Red phosphorus and black phosphorus can be prepared by subjecting white phosphorus to high temperatures and/or high pressures. White phosphorus is extremely reactive -- if left exposed to oxygen in the air, it spontaneously bursts into flame! Thus, it is usually stored under water, with which it does not react.

White phosphorus reacts with excess oxygen to form P_4O_{10}, which reacts with water to form phosphoric acid (H_3PO_4). Phosphoric acid is a relatively weak acid which is added to some carbonated beverages in order to provide extra tartness. Phosphate ions are stored in our bodies as <u>adenosine triphosphate</u> (better known as <u>ATP</u>) which is a useful source of energy for biological processes. Other phosphate compounds make up important components of bones, tooth enamel, and nerve tissues. Because living things need phosphates to survive, phosphates (usually calcium and ammonium phosphates) are used in fertilizers, dentifrices (toothpastes), and baking powders. Some organic phosphates were formerly used as detergents, but their use has been largely discontinued, since they tend to cause pollution of ground waters.

Red phosphorus is used in incendiary devices such as fireworks and safety matches (the red color of the head of a match is due to red phosphorus.). Black phosphorus, the most stable of the three forms, has semiconductor properties.

THE CHEMISTRY OF OXYGEN:

Oxygen was discovered in 1774 by the English chemist Joseph Priestley and the Swedish chemist Carl Wilhelm Scheele, working independently from each other. It is the most abundant element on earth, making up about 21% of the atmosphere, about 50% of the earth's crust, about 67% of the human body, and about 90% of water. Oxygen consists of three major isotopes: oxygen-16 ($^{16}_{8}O$), oxygen-17 ($^{17}_{8}O$), and oxygen-18 ($^{18}_{8}O$). Naturally-occurring oxygen is about 99.8% oxygen-16, with most of the other 0.2% being oxygen-18. Oxygen occurs naturally as one of two allotropes: "regular" oxygen (called dioxygen, or O_2) and ozone (O_3). Molecules of O_2 have an unusual feature -- each oxygen atom carries one unpaired electron. Since molecules having unpaired electrons are called free radicals, O_2 is sometimes referred to as a diradical. (Free radicals are usually harmful to human beings, and it has been said that O_2 would be classified as a hazardous substance if we didn't need it to survive!) Ozone is formed from O_2 by an electrical discharge or ultraviolet light:

$$O_2 + energy \longrightarrow 2\,O. \qquad O + O_2 \longrightarrow O_3.$$

(The "musty" smell in the air after a thunderstorm is due to freshly-formed ozone.) Ozone is important for our survival, since the layer of ozone in the earth's upper atmosphere absorbs harmful ultraviolet light from the sun. In 1974, Professor F. Sherwood Rowland of the University of California at Irvine announced the results of his studies which showed that the chlorofluorocarbons (CFC's) in air conditioners, refrigerators, and aerosol sprays were gradually destroying the earth's ozone layer. Rowland's findings were ridiculed until 1988, but in 1989 the major industrialized nations of the world agreed to stop making CFC's by the year 2000. Ozone can also accumulate in smog in the lower atmosphere, where it can be harmful if inhaled over long periods of time. This is due to the fact that ozone is a powerful oxidizing agent -- it can oxidize the tissues of the lungs if present for long enough.

REACTIONS OF OXYGEN:

Pure oxygen is usually obtained on an industrial scale by cooling air until it liquefies, then distilling it. However, small samples of pure oxygen can be obtained by heating mercuric oxide or potassium chlorate:

$$2\ HgO_{(s)} + heat \longrightarrow 2\ Hg_{(l)} + O_{2\ (g)}.$$

$$2\ KClO_{3\ (s)} + heat \longrightarrow 2\ KCl_{(s)} + 3\ O_{2\ (g)}.$$

Pure oxygen is also obtained industrially by the electrolysis of solutions of KNO_3.

Oxygen reacts with <u>metals</u> to form <u>ionic</u> oxides, which form <u>bases</u> if they react with water. Oxygen reacts with <u>nonmetals</u> to form <u>covalent</u> oxides, which form <u>acids</u> if they react with water. Examples of each case are given below:

$$2\ Mg + O_2 \longrightarrow 2\ MgO. \qquad MgO + H_2O \longrightarrow Mg(OH)_2.$$

$$S_8 + 8\ O_2 \longrightarrow 8\ SO_2. \qquad SO_2 + H_2O \longrightarrow H_2SO_3.$$

The reactions of sulfur (above) illustrate one of the problems of burning sulfur-containing fuels (such as high-sulfur coal) -- namely, the SO_2 gas which is formed goes into the atmosphere, reacts with water vapor, and returns to earth in the form of <u>acid rain</u>. Many factories which emit large amounts of sulfur oxides and nitrogen oxides are located in the northeastern U.S. and southeastern Canada, and these two nations have begun to explore ways to combat acid rain by reducing these emissions.

<u>Peroxides</u> are compounds which contain an oxygen-oxygen single bond. The most common peroxide is hydrogen peroxide (H-O-O-H), which is used in diluted form as a bleach and an antiseptic, and in more concentrated solution as a propellant for small rockets. Benzoyl peroxide (C_7H_5O-O-O-C_7H_5O) is used in anti-acne ointments.

Aside from its obvious use in supporting life, O_2 is used in other ways. Its largest industrial use is in steel blast furnaces -- adding pure O_2 to molten steel oxidizes impurities in the steel. Liquid oxygen is used as a propellant for large rockets. Large quantities of O_2 are also used for oxy-acetylene welding.

THE CHEMISTRY OF SULFUR:

Sulfur occurs naturally both in its elemental form and in many minerals. There are four naturally-occurring isotopes of sulfur: sulfur-32 ($^{32}_{16}S$), sulfur-33 ($^{33}_{16}S$), sulfur-34 ($^{34}_{16}S$), and sulfur-36 ($^{36}_{16}S$). Sulfur-32 makes up about 95.0% of naturally-occurring sulfur, with sulfur-34 (4.2%) and sulfur-33 (0.8%) making up most of the remainder. Sulfur exists in many different allotropic forms, of which the most stable is "sulfur-eight" (S_8). In this form, molecules of sulfur consist of eight sulfur atoms arranged in a crown-shaped ring. Elemental sulfur can be obtained from mines along the Texas and Louisiana coasts by the <u>Frasch Process</u>, in which hot water is forced into the sulfur deposit to melt the sulfur and high-pressure air is used to force the molten sulfur to the surface. Sulfur is used in the process called <u>vulcanization</u>, in which it is allowed to react with rubber to strengthen the rubber for use in automobile tires. It is also used in gunpowder.

Sulfur is also used as a raw material for the manufacture of its oxides and products derived from them. The most important of these are sulfuric acid (H_2SO_4) and sulfurous acid (H_2SO_3), which can be prepared as shown below:

$$S_8 + 8\ O_2 \longrightarrow 8\ SO_2. \qquad SO_2 + H_2O \longrightarrow H_2SO_3.$$
$$2\ SO_2 + O_2 \longrightarrow 2\ SO_3. \qquad SO_3 + H_2O \longrightarrow H_2SO_4.$$

When these reactions occur in the atmosphere, they generate acids which return to earth in the form of <u>acid rain</u>. However, these reactions can be used in industry for beneficial purposes. For example, sulfuric acid is used in large quantities to make fertilizers such as ammonium sulfate, and salts of sulfurous acid (known as <u>sulfites</u>) have been used as preservatives in fruits and vegetables.

Other sulfur compounds include hydrogen sulfide (which is added to natural gas to provide the aroma of "rotten eggs") and several amino acids (bonds between sulfur atoms help give some proteins their necessary geometries).

THE CHEMISTRY OF FLUORINE:

Elemental fluorine was first isolated in 1886 by the French chemist Henri Moissan, who won the Nobel Prize in Chemistry in 1906 for this achievement. Moissan used electrochemical methods to isolate pure fluorine: $2 F^- \longrightarrow F_2 + 2 e^-$. Because fluorine is the most reactive and most electronegative of all the elements, it was thought for many years that Moissan's method was the only way to prepare elemental fluorine, but Rockwell International's Karl Christe showed in 1986 that elemental fluorine could be prepared using the oxidation-reduction reaction below:

$$2 MnF_6^{2-} + 4 SbF_5 \longrightarrow F_2 + 4 SbF_6^- + 2 MnF_3.$$

Only one isotope of fluorine, fluorine-19 ($^{19}_{9}F$), occurs naturally.

Fluorine occurs naturally in the form of various minerals, among them fluorite (CaF_2) and cryolite (Na_3AlF_6). Fluorine reacts with almost anything, even water: $2 F_2 + H_2O \longrightarrow 2 HF + OF_2$. (Notice that in this reaction, the oxidation state of oxygen changes from -2 to $+2$. Hence, since oxygen is oxidized, fluorine is one of the few substances that can be said to make water "burn"!) The acid formed in this reaction, HF, is a relatively weak acid, but there are two substances it attacks strongly. One of them is human flesh, so chemists working with HF wear many layers of protective clothing. The other is glass (SiO_2, mostly):

$$4 HF + SiO_2 \longrightarrow 2 H_2O + SiF_4.$$

Thus, HF cannot be stored in glass bottles. It is used in the etching of glass.

A familiar fluorine-containing compound is Teflon, which is short for polytetrafluoroethylene. Many molecules of tetrafluoroethylene ($CF_2=CF_2$) can be made into one long molecule (called a polymer) by the process called polymerization:

$$n\ CF_2=CF_2 \longrightarrow \ldots-CF_2-CF_2-CF_2-CF_2-CF_2-CF_2-\ldots = (CF_2-CF_2)_n.$$

Teflon is used as a "non-stick" coating for cooking utensils.

Fluoride ions are added to drinking water to reduce dental cavities.

THE CHEMISTRY OF CHLORINE:

Chlorine was discovered in 1774 by the Swedish chemist Carl Wilhelm Scheele. The two major naturally-occurring isotopes of chlorine are chlorine-35 ($^{35}_{17}Cl$) and chlorine-37 ($^{37}_{17}Cl$). Chlorine-35 makes up 75.53% of naturally-occurring chlorine, with chlorine-37 making up the other 24.47%. This unusual 3:1 ratio of the two isotopes accounts for the fact that chlorine's atomic weight is 35.453 amu -- that is, closer to a half-integer value than to an integer value. Elemental chlorine does not occur naturally, but chlorine is found naturally in many minerals. For example, chlorine in the form of sodium chloride (NaCl) makes up about 1.9% of seawater. Elemental chlorine is produced commercially by the electrolysis of sea water or molten sodium chloride. The "half-reaction" is: $2 Cl^- \longrightarrow Cl_2 + 2 e^-$.

The principal use of chlorine is in the manufacture of hypochlorite (OCl^-) salts. This is done by allowing chlorine to react with bases, as shown:

$$Ca(OH)_2 + 2 Cl_2 \longrightarrow Ca(OCl)_2 + 2 HCl.$$
$$NaOH + Cl_2 \longrightarrow NaOCl + HCl.$$

Calcium hypochlorite ($Ca(OCl)_2$) is used as a bleaching powder for many different kinds of fabrics. Sodium hypochlorite (NaOCl) is used to purify and disinfect water -- it's the "chlorine" that is added to swimming pools. It is also the active ingredient in liquid "chlorine bleach". (Notice, by the way, that the reactions above are equilibria. Adding acid to household bleach drives the equilibrium to the left, forming elemental chlorine, which is irritating to the eyes, nose, and lungs.)

Other chlorine-containing compounds include hydrochloric acid (HCl, also known as "stomach acid"!), vinyl chloride (C_2H_3Cl, which can be polymerized to make polyvinyl chloride, a plastic used in raincoats and upholstery), some insecticides (for example, DDT stands for "dichlorodiphenyltrichloroethane"), and the ozone-consuming compounds known as chlorofluorocarbons (also known as Freons).

THE CHEMISTRY OF BROMINE AND IODINE:

Bromine was discovered by the French chemist Antoine-Jerome Balard in 1826. The two major naturally-occurring isotopes of bromine are bromine-79 ($^{79}_{35}Br$) and bromine-81 ($^{81}_{35}Br$). Bromine-79 makes up 50.54% of naturally-occurring bromine, with bromine-81 making up the other 49.46%. The 1:1 ratio of the two isotopes leads to the unusual situation of bromine having an atomic weight of 79.904 amu (approximately 80 amu) despite the fact that no individual bromine atoms have a mass of 80 amu! Elemental bromine can be prepared by the reaction of bromide ions in sea water with chlorine: $Cl_2 + 2 Br^- \longrightarrow Br_2 + 2 Cl^-$. Bromine is the only nonmetal that is a liquid. The major use of bromine is in the preparation of ethylene dibromide ($C_2H_4Br_2$), which is added to leaded gasoline to prevent knocking. Bromine is also useful as an analytical reagent. For example, it can be used to determine the degree of "unsaturation" in "polyunsaturated" fats and oils.

Iodine was discovered by the French chemist Bernard Courtois in 1811. Only one isotope of iodine, iodine-127 ($^{127}_{53}I$), occurs naturally. Iodine occurs naturally in the form of iodide salts, which can be isolated from seaweed and from mineral deposits. Elemental iodine can be prepared by the following redox reaction:

$$4 KI + 2 CuSO_4 \cdot 5H_2O \longrightarrow I_2 + 2 CuI + 2 K_2SO_4 + 10 H_2O.$$

Iodine is a purple crystalline solid which sublimes instead of melting. Iodine is used medicinally in many ways. The "iodine" that is used as an antiseptic is really a solution of iodine in ethyl alcohol (sometimes called "tincture of iodine"). Our bodies need iodine in order to make the hormone <u>thyroxin</u>, which is produced by the thyroid gland to stimulate the human metabolism. Iodine can be consumed in the form of "iodized salt" (NaCl with KI added); lack of iodine in the diet causes swelling of the thyroid gland (known as a <u>goiter</u> condition). Some thyroid conditions are treated by ingesting iodine-131 ($^{131}_{53}I$), a radioactive isotope of iodine.

THE CHEMISTRY OF THE NOBLE GASES:

The chemistry of the noble gases can be summarized in about two words: "not much"! They're called the "noble" gases because they tend not to combine with the "common" elements to form compounds. The following table lists the noble gases and some of the ways in which each is commonly used.

Helium is the second most abundant element in the universe, since it is formed from the fusion of hydrogen atoms in stars. Its low flammability and low density make it ideal for use in "lighter-than-air" balloons and dirigibles. It is also used in place of nitrogen in the "air" in SCUBA tanks, since it is insoluble in the bloodstream and thus unlikely to cause the "bends" upon decompression.

Neon is familiar as the gas used to make "neon" lights. These are gas discharge tubes filled with neon gas at low pressure. When excited by electricity, neon atoms emit a reddish-orange glow. Other noble gases can also be used in this way, each generating a different color -- helium tubes give off a red-violet glow, argon tubes are purple, krypton tubes are violet, and xenon tubes are blue-green.

Argon is used as an inert gas shield in arc welding, in certain kinds of LASERs, and to provide an inert atmosphere when nitrogen cannot be used.

Krypton is used in certain kinds of incandescent and fluorescent light bulbs to provide an inert atmosphere, thus keeping oxygen away from the filament and preventing the filament from burning out. One compound, KrF_2, is known.

Xenon is used as an anaesthetic and in high-brilliance lamps designed to resemble natural daylight. Compounds include XeF_2, XeF_4, XeF_6, and XeO_3.

Radon is a radioactive gas formed by the radioactive decay of radium, thorium, and actinium in the earth's crust. In the 1980's, the accumulation of radon in the basements of homes has been a matter of public concern. Radon is used as a source of the radiation used to treat certain kinds of cancer.

THE CHEMISTRY OF THE ALKALI METALS:

The alkali metals are located in Group IA of the Periodic Table. Thus, their atoms each have <u>one</u> valence electron. The alkali metals are among the most reactive of the metals, readily losing their valence electrons to form ions with a <u>+1</u> charge. The "half-reaction" is: $M \longrightarrow M^+ + e^-$. (Here, "M" is used to represent the symbol for <u>any</u> of the alkali metals.) For example, the alkali metals react vigorously with water, forming hydrogen gas and the hydroxide of the metal: $2 M + 2 H_2O \longrightarrow H_2 + 2 MOH$. The relative reactivity of the alkali metals increases as their atomic numbers increase -- potassium is more reactive than sodium, which in turn is more reactive than lithium. The alkali metals tend to be relatively soft metals (lithium has a density of 0.534 g/mL, making it the lightest of all the metals), with relatively low melting points (sodium melts at 97.82 ^{O}C). Since they are so reactive, the alkali metals do not occur in nature in elemental form, but can be prepared by electrochemical methods: $M^+ + e^- \longrightarrow M$. For example, sodium is prepared from molten sodium chloride in a <u>Downs cell</u>.

Some alkali metal compounds and their common uses are listed below:

<u>Lithium</u> compounds produce a bright red color when placed in a burner flame. Lithium carbonate (Li_2CO_3) has been used to treat manic-depressive syndrome.

<u>Sodium</u> compounds produce a bright yellowish-orange color when placed in a burner flame. For this reason, sodium oxalate ($Na_2C_2O_4$) and cryolite (Na_3AlF_6) are used to produce yellow bursts in fireworks displays.[1] Other common compounds of sodium include sodium chloride (NaCl, table salt), sodium hydroxide (NaOH, used in making soap), and sodium bicarbonate ($NaHCO_3$, "baking soda").

<u>Potassium</u> compounds produce a faint violet color when placed in a flame from a burner. Potassium nitrate (KNO_3) is used in gunpowder and in fireworks.[1]

---

1) Conkling, J. A., "Chemistry of Fireworks", <u>Chemical and Engineering News</u>, June 29, 1981, page 27.

THE CHEMISTRY OF THE ALKALINE EARTH METALS:

The alkaline earth metals are located in Group IIA of the Periodic Table. Thus, their atoms each have <u>two</u> valence electrons. The alkaline earth metals react by losing their valence electrons to form ions with a $\underline{+2}$ charge, as shown by the "generic" reaction: $M \longrightarrow M^{2+} + 2\ e^-$. The relative reactivity of the alkaline earth metals increases as their atomic numbers increase -- for example, calcium reacts with water ($Ca + 2\ H_2O \longrightarrow Ca(OH)_2 + H_2$), whereas magnesium does not. The alkaline earth metals do not occur naturally in elemental form, but can be prepared by electrochemical methods. For example, magnesium is prepared from magnesium chloride in the <u>Dow Process</u>: $MgCl_2 \longrightarrow Mg + Cl_2$.

Some alkaline earth metal compounds and their uses are listed below:

<u>Beryllium</u> is named for the mineral beryl ($Be_3Al_2Si_6O_{18}$), which is more familiar as the gemstones emerald and aquamarine.

<u>Magnesium</u> produces a brilliant white light when it burns, and is used in fireworks to produces flashes of bright white light.[1] Its compounds include magnesium hydroxide ($Mg(OH)_2$, "milk of magnesia") and the chlorophylls in plants.

<u>Calcium</u> compounds produce a reddish-orange color when placed in a flame from a burner. Common calcium compounds include calcium carbonate ($CaCO_3$, found in limestone, marble, chalk, and eggshells), calcium sulfate dihydrate ($CaSO_4 \cdot 2\ H_2O$, "plaster of Paris"), and fluoroapatite ($Ca_{10}(PO_4)_6F_2$, found in tooth enamel).

<u>Strontium</u> compounds produce a bright red color when placed in a burner flame. For this reason, strontium carbonate ($SrCO_3$) and strontium nitrate ($Sr(NO_3)_2$) are used to produce red bursts in fireworks displays.[1]

<u>Barium</u> compounds produce a green color when placed in a burner flame.[1] Barium sulfate ($BaSO_4$) is used in X-ray diagnostic work as a contrast medium.

---

1) Conkling, J. A., "Chemistry of Fireworks", <u>Chemical and Engineering News</u>, June 29, 1981, page 27.

AN INTRODUCTION TO THE TRANSITION METALS:

The transition metals are located in the "B" groups in the center of the Periodic Table, as well as in the two rows usually placed at the bottom of the Periodic Table. (The elements in the "B" groups are sometimes called the "d-block" elements, because their atoms contain partially-filled d subshells. Similarly, the elements in the bottom two rows (the lanthanides and the actinides) are sometimes called the "f-block" elements, because their atoms contain partly-full f subshells.)

There are two properties that distinguish transition metals from alkali metals or alkaline earth metals. One is that atoms of transition metals may lose different numbers of electrons, forming stable ions in more than one oxidation state. For example, copper forms both Cu^+ and Cu^{2+} ions, and iron forms both Fe^{2+} and Fe^{3+} ions. (By contrast, alkali metals form only M^+ ions, and alkaline earth metals form only M^{2+} ions.) The other is that transition metal ions tend to form complex ions. A complex ion (sometimes called a coordination complex) is formed when one or more ligands are joined to a central metal ion by covalent bonds. A ligand is an anion or a neutral molecule which donates one or more electron pairs to a metal ion in the process of forming a covalent bond to it. (Since both electrons in this kind of covalent bond come from the ligand, these are coordinate covalent bonds, hence the name "coordination complex".) Examples of ligands are Cl^- and CN^- ions and the neutral molecules NH_3 and H_2O. An example of a complex ion is the ion formed when two molecules of NH_3 become attached to a silver ion:

$$Ag^+ \text{ (aq)} + 2 \text{ } NH_3 \text{ (aq)} \rightleftharpoons Ag(NH_3)_2^+ \text{ (aq)}.$$

The equilibrium above is described by an equilibrium constant, K_{form} (sometimes called the formation constant for the complex ion in question). For the complex ion above, $K_{form} = [Ag(NH_3)_2^+]/[Ag^+][NH_3]^2 = 1.4 \times 10^7$.[1]

---

1) From "Fundamentals of Analytical Chemistry" by D. A. Skoog and D. M. West. Third Edition, 1976 by Holt, Rinehart, and Winston, page 786.

84

USING FORMATION CONSTANTS:

Problem: Calculate the solubility of AgCl in a 1.00 M solution of NH_3. The equilibria are: $AgCl_{(s)} \rightleftharpoons Ag^+_{(aq)} + Cl^-_{(aq)}$. $K_{sp} = 1.82 \times 10^{-10}$.[1] $Ag^+_{(aq)} + 2 NH_{3(aq)} \rightleftharpoons Ag(NH_3)_2^+_{(aq)}$. $K_{form} = 1.4 \times 10^7$.[1]

Solution: The large value of K_{form} implies that virtually all of the Ag^+ ions in solution are converted into complex ions. Thus, the real equilibrium is: $AgCl_{(s)} + 2 NH_{3(aq)} \rightleftharpoons Ag(NH_3)_2^+_{(aq)} + Cl^-_{(aq)}$. The equilibrium expression for this heterogeneous equilibrium is: $K_{eq} = [Ag(NH_3)_2^+][Cl^-]/[NH_3]^2$. We don't know the value of K_{eq}, but we can find it by multiplying K_{sp} by K_{form}:

$$K_{sp} = [Ag^+][Cl^-]. \qquad K_{form} = [Ag(NH_3)_2^+]/[Ag^+][NH_3]^2.$$

$$K_{sp} \times K_{form} = ([Ag^+][Cl^-])([Ag(NH_3)_2^+]/[Ag^+][NH_3]^2).$$

$$K_{sp} \times K_{form} = [Cl^-][Ag(NH_3)_2^+]/[NH_3]^2 = K_{eq}.$$

$$K_{eq} = K_{sp} \times K_{form} = (1.82 \times 10^{-10})(1.4 \times 10^7) = \underline{2.5 \times 10^{-3}}.$$

Now that K_{eq} is known, this problem can be solved by the usual equilibrium methods:

	Initial Value	Change During Reaction	Final Value
$[Ag(NH_3)_2^+]$	0.00 M	+X	X M
$[Cl^-]$	0.00 M	+X	X M
$[NH_3]$	1.00 M	-2X	(1.00 - 2X) M

$$K_{eq} = 2.5 \times 10^{-3} = [Ag(NH_3)_2^+][Cl^-]/[NH_3]^2 = X^2/(1.00 - 2X)^2.$$

Taking the square root of both sides allows us to solve this equation more easily:

$$X/(1.00 - 2X) = (2.5 \times 10^{-3})^{1/2} = 5.0 \times 10^{-2} = 0.050.$$

$$X = (0.050)(1.00 - 2X) = 0.050 - (0.050)(2X) = 0.050 - (0.10)X.$$

$$X + (0.10)X = 0.050 = (1.00 + 0.10)X = (1.10)X.$$

$$X = 0.050/1.10 = \underline{0.045} = [Ag(NH_3)_2^+] = [Cl^-] = \text{solubility}.$$

Notice that this is greater than the 1.3×10^{-5} M predicted for AgCl in pure water.

---

1) From "Fundamentals of Analytical Chemistry" by D. A. Skoog and D. M. West. Third Edition, 1976 by Holt, Rinehart, and Winston, pages 783-786.

THE CHEMISTRY OF THE TRANSITION METALS IN GROUPS IIIB AND IVB:

Scandium is costly to produce, so it isn't used widely yet. However, its low density and high melting point (1541 $^{\circ}$C) suggests a potential use in space missiles as a structural metal.

Yttrium was thought to have limited use until it was discovered in 1987 that mixed oxides of yttrium (for example, $YBa_2Cu_3O_n$) behave as superconductors at relatively high temperatures. Superconductivity is the ability of a substance to conduct an electrical current with no electrical resistance at all. Until 1987, it was thought that superconductivity could only occur at extremely low temperatures (about 4 K, the temperature of liquid helium), but the yttrium compound above is a superconducting material at temperatures as "high" as 90 K (-183 $^{\circ}$C, slightly above the temperature of liquid nitrogen). Possible commercial applications of this phenomenon are the subject of ongoing research.

Titanium is as strong as steel, but less dense. These properties make it useful as a structural metal in aircraft. Important titanium compounds include titanium tetrachloride ($TiCl_4$), which catalyzes certain polymerization reactions and reacts with water to form HCl gas and a thick, white smoke of TiO_2:

$$TiCl_4 + 2\ H_2O \longrightarrow TiO_2 + 4\ HCl.$$

These properties made $TiCl_4$ useful in making "smoke bombs" during World War II. Titanium dioxide (TiO_2) is used as a white pigment in house paints.

Zirconium occurs naturally in the mineral zircon, which is present in a number of gemstones. Zircon is essentially zirconium silicate ($ZrSiO_4$), which has been used as an abrasive in toothpastes.

Hafnium atoms have a strong tendency to absorb neutrons. This property makes hafnium useful in the "control rods" which regulate the rate at which "chain reactions" occur in nuclear reactors.

THE CHEMISTRY OF THE TRANSITION METALS IN GROUPS VB AND VIB:

Vanadium is added to steel to make it more resistant to rusting. An oxide of vanadium, vanadium(V) oxide (V_2O_5), is used as a catalyst in the "contact process" which converts SO_2 into SO_3 during the commercial production of H_2SO_4.

Niobium has been alloyed with zirconium to make superconducting magnets. It is hoped that this property will allow the generation of electric power on a large scale. Niobium is also used in stainless steel and in nuclear reactors.

Tantalum is almost completely unreactive at temperatures below 150 $^{\circ}$C. This property allows it to be used in place of human bone in patients who have suffered severe injuries. Tantalum has also been used in pen points and aircraft.

Chromium derives its name from the Greek word "chroma", meaning "color". Many of the compounds of chromium are very colorful. For example, lead chromate ($PbCrO_4$) is a yellow pigment used in colored chalk, and chromium(VI) oxide (Cr_2O_3) is a green pigment used in roofing materials. Various coordination complexes of chromium are also colorful, and the color can be varied simply by changing the identities of the ligands used to make the complexes. Compounds which contain chromium in its +6 oxidation state are usually good oxidizing agents; for example, sodium dichromate ($Na_2Cr_2O_7$) is used in the tanning industry to react with the collagen in animal hides, converting the hides into hard, tough leather. Chromium metal is used to make stainless steel and the "chrome" finishes on automobiles.

Molybdenum is an essential trace element in plant nutrition, since it is present in the enzyme which catalyzes the process of nitrogen fixation. This process converts nitrogen gas into ammonia, which plants need to survive.

Tungsten has the highest melting point of all the elements -- 3410 $^{\circ}$C! This property makes it useful in incandescent light bulbs as the glowing filaments. Tungsten is added to steel to harden it for high-speed tools such as dental drills.

THE CHEMISTRY OF THE TRANSITION METALS IN GROUPS VIIB AND VIIIB:

Manganese occurs naturally in nodules found at the bottoms of oceans and several of the Great Lakes. It is used to make steel tougher and more flexible. Manganese dioxide (MnO_2) and potassium permanganate ($KMnO_4$) are oxidizing agents.

Technetium gets its name from the Greek word "technetos", which means "artificial". Technetium atoms do not occur naturally; they were first prepared by the nuclear bombardment of molybdenum atoms. Coordination complexes of radioactive technetium-99m ($^{99m}_{43}Tc$; the "m" stands for "metastable") have applications in nuclear medicine; specific tissues can be targeted by judicious choices of ligands.[1]

Rhenium has a very high melting point (3180 $^{\circ}C$), which makes it useful in the manufacturing of thermocouples. Thermocouples are devices used to measure temperatures up to 2200 $^{\circ}C$. Rhenium wire is also used to make filaments.

Iron is the most abundant and inexpensive of all the metals. It makes up about 5% of the earth's crust and the vast majority of the earth's molten core. It occurs naturally as the ore hematite (Fe_2O_3), from which elemental iron can be obtained by reduction with carbon (coke) in a blast furnace. The resulting iron is called "pig iron", which contains about 4% carbon. Pure iron can be refined using the basic oxygen process, in which pure oxygen is passed through molten pig iron to oxidize the impurities. Iron is the major component of steel. It is present in blood (hemoglobin) and in "Prussian Blue" ($Fe_4[Fe(CN)_6]_3 \cdot 16H_2O$), a blue dye.

Ruthenium is a hard and relatively unreacted metal. It is alloyed with palladium and platinum to make hard, wear-resistant electrical contacts.

Osmium has the greatest density of all the elements -- 22.48 g/mL! It is used to produce very hard alloys -- pen points and phonograph needles are about 60% osmium. Osmium tetroxide (OsO_4) is a useful oxidizing agent in organic labs.

---

1) From a lecture given at the University of Delaware on November 5, 1980, by Dr. Edward A. Deutsch, Department of Chemistry, University of Cincinnati.

THE CHEMISTRY OF THE TRANSITION METALS IN GROUP VIIIB:

Cobalt compounds are usually very colorful, and have been used to give color to porcelains, tiles, and enamels. Cobalt ions are present in Vitamin B_{12}. In this molecule, the cobalt ions are covalently bonded to each of four nitrogen atoms which are located in a large, cyclic organic molecule. The cyclic organic molecule is called a porphyrin, and Vitamin B_{12} is an example of a metalloporphyrin (a porphyrin which contains a metal ion). Other examples of metalloporphyrins are hemoglobin (iron ions) and chlorophyll (magnesium ions). Porphyrins are examples of polydentate ligands, since more than one atom in the ligand binds to the metal.

Rhodium is highly reflective and relatively unreactive. It is used in making mirrors, and is alloyed with other metals to protect them against corrosion.

Iridium gets its name from the Latin "iris", meaning "rainbow", since its compounds are usually colorful. Its high density (22.42 g/mL) makes it useful in fountain pen points and in hardening certain alloys.

Nickel is used to make nickels! (Nickels are about 25% nickel.) It is also used in making stainless steel and "nichrome" (nickel-chromium) wire. Nickel is an important catalyst for hydrogenation reactions in organic chemistry.

Palladium is alloyed with gold to make "white gold", which is used in jewelry and in dental fillings. It has a strong tendency to absorb hydrogen, which makes it useful as a hydrogenation catalyst.

Platinum is similar to nickel and palladium in its chemistry. One important platinum compound is the coordination complex shown at the right, which has been effective in treating certain kinds of cancer. Notice the difference between the cis compound above and the trans compound below: the atoms are the same, but their spatial arrangement is different. Compounds which differ in this way are called stereoisomers.

THE CHEMISTRY OF THE TRANSITION METALS IN GROUPS IB AND IIB:

Copper occurs naturally as its sulfide ores, which are converted to the oxides by heating them in the presence of oxygen in a process called roasting:

$$2 \ Cu_2S \ + \ 3 \ O_2 \ \longrightarrow \ 2 \ Cu_2O \ + \ 2 \ SO_2.$$

The oxides can be converted to the pure metal by heating them in the presence of carbon (in the form of coke) in the process known as smelting:

$$2 \ Cu_2O \ + \ C \ \longrightarrow \ 4 \ Cu \ + \ CO_2.$$

Copper conducts electricity well, and is thus used to make wires. It is alloyed with zinc to make brass and with zinc and tin to make bronze. (Pennies are bronze.)

Silver conducts electricity better than copper does, but silver is more expensive. It is used in making coins, jewelry, and "silverware" (actually steel with a silver coating). Silver halides become gray upon exposure to light, which makes them useful in photography. The "tarnish" on silverware is silver sulfide.

Gold occurs naturally in elemental form, and is obtained by "panning" for gold nuggets and separating out the impurities. It is used in making jewelry and as a standard for currency. Some gold compounds are used to treat arthritis.

Zinc is a good reducing agent, and is thus used in dry cell batteries and to inhibit corrosion of other metals. Important zinc compounds include zinc oxide (ZnO, used in sun screens) and zinc sulfide (ZnS, used in television screens).

Cadmium atoms have a strong tendency to absorb neutrons, which makes cadmium useful in making the "control rods" which regulate the rate at which "chain reactions" occur in nuclear reactors. Cadmium sulfide (CdS) is a yellow pigment.

Mercury is one of the few metals which is a liquid at room temperature. It is used in thermometers and in "mercury arc lamps" which are used to illuminate streets at night. Alloys of mercury with other metals are called amalgams and are usually soft and malleable; silver amalgam has been used in dental fillings.

THE CHEMISTRY OF THE GROUP IIIA METALS:

<u>Aluminum</u> occurs naturally in the form of minerals such as <u>bauxite</u> (Al_2O_3) and <u>cryolite</u> (Na_3AlF_6). In 1886, a 21-year-old Oberlin College student named Charles Hall became the first person to prepare metallic aluminum by passing a current of electricity through a molten mixture of bauxite and cryolite. The Hall Process is still used to manufacture aluminum, but it is rather costly -- a better way to obtain aluminum now is to recycle aluminum cans! Metallic aluminum is used in cans, "aluminum foil", kitchen utensils, and "aluminum siding" for homes. Its low density and high malleability make it useful as a structural metal for aircraft. Powdered aluminum is used as a fuel in pyrotechnic mixtures such as fireworks[1] and the fuel used in the solid booster rockets in space shuttles. It has also been used for spot-welding in the Thermite Reaction: $2\ Al\ +\ Fe_2O_3\ \longrightarrow\ Al_2O_3\ +\ 2\ Fe$. Aluminum compounds of interest include alumina (Al_2O_3, found in gemstones such as rubies and sapphires), beryl ($Be_3Al_2Si_6O_{18}$, found in emeralds), aluminum chloride hexahydrate ($AlCl_3 \cdot 6H_2O$, called "aluminum chlorohydrate" when used in deodorants), and <u>alums</u> such as $KAl(SO_4)_2 \cdot 12H_2O$, which is used medicinally as an astringent.

<u>Gallium</u> is a liquid on warm days -- its melting point is 30 ^oC! This property makes it useful in thermometers designed to measure high temperatures. Gallium arsenide (GaAs) is able to convert electricity directly into light, which makes it useful in transistors and solar energy cells.

<u>Indium</u> is similar to gallium in its chemistry. It is used in alloys for making bearings, in transistors with germanium, and in solar cells.

<u>Thallium</u> and its compounds are quite toxic; thallium sulfate (Tl_2SO_4) is used commercially as a rat poison. One isotope of thallium, thallium-201 ($^{201}_{81}Tl$), is radioactive and is used in coronary artery medicine as an imaging agent.

---

1) Conkling, J. A., "Chemistry of Fireworks", <u>Chemical and Engineering News</u>, June 29, 1981, page 27.

THE CHEMISTRY OF THE GROUP IVA AND GROUP VA METALS:

<u>Germanium</u> was used to make the first transistors. Its semiconductor properties make it useful in the manufacture of electronic devices such as diodes. Some germanium compounds are useful in fighting certain forms of bacteria.

<u>Tin</u> is used in making "tin cans" (actually steel cans covered with a thin coating of tin) because it doesn't rust and resists corrosion. Tin is alloyed with zinc and copper to make <u>bronze</u>. It occurs naturally as the mineral cassiterite (SnO_2), from which metallic tin can be isolated by reduction with carbon (smelting):

$$SnO_2 + C \longrightarrow Sn + CO_2.$$

Some of the more interesting compounds of tin include stannous chloride ($SnCl_2$, a good reducing agent used in the manufacture of dyes) and stannous fluoride (SnF_2, the "Fluoristan" added to toothpaste as a source of fluoride to prevent cavities).

<u>Lead</u> has been used for centuries to make pipes for plumbing (the words "plumbing" and "plumber" come from the Latin word "plumbum", which is the origin of the symbol "Pb" for lead). Lead occurs naturally as the mineral <u>galena</u> (PbS); the element itself can be isolated by roasting (reaction with O_2) followed by smelting:

$$2\ PbS + 3\ O_2 \longrightarrow 2\ PbO + 2\ SO_2.$$
$$2\ PbO + C \longrightarrow 2\ Pb + CO_2.$$

Important lead compounds include tetraethyllead ($Pb(C_2H_5)_4$, used as an "anti-knock" agent in "leaded" gasoline), lead arsenate ($PbHAsO_4$, used as an insecticide), and lead chromate ($PbCrO_4$, used as a yellow pigment in paints and chalk).

<u>Antimony</u> is alloyed with lead to make it harder. Its compounds are used in manufacturing paints, ceramic enamels, pottery, and fireproofing materials.

<u>Bismuth</u> has a relatively low melting point (271 $^\circ C$), and is thus used to make low-melting alloys for soldering and automatic sprinkler systems. Bismuth oxychloride (BiOCl) is used in face powders; "Pepto-<u>Bismol</u>" is made from $BiC_7H_5O_4$.

NUCLEAR CHEMISTRY:

In the latter part of the twentieth century, the word "nuclear" has come to be associated with catastrophic events, such as "nuclear war" or "nuclear winter". This is unfortunate, because the word "nuclear" is derived from the word "nucleus", and is intended simply to describe things associated with atomic nuclei. Therefore, "nuclear chemistry" is simply the chemistry of the atomic nucleus. Most of the chemical reactions that atoms undergo involve the transfer or sharing of one or more of the atoms' valence electrons, but this is not the case for nuclear reactions, which involve only the nuclei of atoms.

One unusual feature of most nuclear reactions is that atoms of one element are often changed into atoms of another element. The transformation of an atom of one element into an atom of another element is called transmutation. One way in which transmutation can occur is for an unstable atomic nucleus to stabilize itself by giving off radiation in the form of small particles of matter or photons of energy. An atomic nucleus which gives off radiation of this type is said to be radioactive. The radioactivity of an atomic nucleus is its relative tendency to give off this kind of radiation. ("Radioactivity" also means the radiation itself.)

Some of the particles of matter typically emitted by radioactive nuclei are listed below. The emission process is sometimes called "radioactive decay", since a large atomic nucleus "decays" in the process of forming smaller nuclei.

Name of Particle	Symbol(s) of Particle	Description
Alpha Particle	$^{4}_{2}He$, $^{4}_{2}He^{2+}$, or $^{4}_{2}\alpha$	helium nucleus
Beta Particle	$^{0}_{-1}e$, $^{0}_{-1}e^{-}$, or $^{0}_{-1}\beta$	electron
Gamma Particle	$^{0}_{0}\gamma$	photon of energy
Positron	$^{0}_{1}e$, $^{0}_{1}e^{+}$, or $^{0}_{1}\beta^{+}$	positive electron
Neutron	$^{1}_{0}n$	neutron

BALANCING EQUATIONS OF NUCLEAR REACTIONS:

The total mass of the products of a nuclear reaction is equal to the total mass of the reactants. Similarly, the total charge of the products of a nuclear reaction is equal to the total charge of the reactants. Therefore, to write a balanced equation for a nuclear reaction, all that is necessary is to be sure that the sum of the mass numbers (superscripts) is the same for the reactants as it is for the products, and that the sum of the atomic numbers (subscripts) is the same for the reactants as it is for the products. Examples are given below.

Problem: A uranium-238 nucleus undergoes radioactive decay, becoming a nucleus of thorium-234. Write a balanced equation for this nuclear reaction.

Solution: The difference between the mass numbers of the reactant and product is 4 (238 - 234). Hence, it is likely that the emitted particle is an alpha particle, since an alpha particle has a mass number of 4. This is confirmed by the fact that the difference in atomic numbers between uranium (atomic number = 92) and thorium (atomic number = 90) is 2 (92 - 90). This agrees with the atomic number of an alpha particle (a helium nucleus), which is also 2. Therefore, the balanced equation for this nuclear reaction is: $^{238}_{92}U \longrightarrow {}^{234}_{90}Th + {}^{4}_{2}\alpha$.

Problem: A thorium-234 nucleus undergoes radioactive decay, giving off a beta particle and a gamma particle. Write a balanced equation for this reaction.

Solution: Beta particles and gamma particles both have mass numbers of zero, so no change will occur in the atom's mass number -- it will remain at 234. Gamma particles have atomic numbers of zero, but the "atomic number" of a beta particle is -1. (Beta particles are electrons, each of which carries a -1 charge.) Therefore, the atomic number of the product nucleus must be the difference between the atomic numbers of thorium (90) and a beta particle (-1). 90 - (-1) = 91, the atomic number of protactinium. The equation is: $^{234}_{90}Th \longrightarrow {}^{234}_{91}Pa + {}^{0}_{-1}\beta + {}^{0}_{0}\gamma$.

NUCLEAR STABILITY:

Atomic nuclei which are radioactive give off energy in order to become more stable. The stability of atomic nuclei depends upon several factors. Since atomic nuclei contain positively-charged protons which repel each other, neutrons stabilize the nucleus by separating the protons and "insulating" them from each other. The greater the number of protons present in an atomic nucleus, the greater the number of neutrons needed for the nucleus to be stable. The ratio of protons to neutrons in stable nuclei is about 1:1 for relatively small nuclei, but is about 2:3 for larger, more massive nuclei. In addition, nuclei that contain <u>even</u> numbers of protons or neutrons are generally more stable than nuclei that contain <u>odd</u> numbers of protons or neutrons. (Of the 264 known stable nuclei, 157 contain <u>even</u> numbers of protons <u>and</u> neutrons, but only 5 contain <u>odd</u> numbers of protons and neutrons.) Also, nuclei which contain certain "magic numbers" of protons or neutrons tend to be more stable than other nuclei. The "magic numbers" are 2, 8, 20, 50, 82, and 126. Nobody knows why these numbers are "magic" for atomic nuclei.

Atomic nuclei which contain too many protons <u>and</u> neutrons to be stable often emit <u>alpha</u> particles. Alpha particles contain two protons and two neutrons each, so emission of an alpha particle considerably reduces the mass of a nucleus. Atomic nuclei which contain too many neutrons to be stable usually emit <u>beta</u> particles. Beta particles are electrons, but there are no electrons in an atomic nucleus. However, when a nucleus emits a beta particle, one neutron in the nucleus is changed into a proton: $^1_0n \longrightarrow \ ^{\ 0}_{-1}\beta + \ ^1_1H$. (The symbol for a proton is ^1_1H.) Atomic nuclei which contain too many protons to be stable usually emit <u>positrons</u>. Positrons are particles the size of electrons, but having a <u>positive</u> charge. When a positron collides with an electron, <u>both</u> are destroyed. Emission of a positron changes a proton in the nucleus into a neutron: $^1_1H \longrightarrow \ ^0_1e^+ + \ ^1_0n$.

BINDING ENERGY:

One way to measure the relative stability of an atomic nucleus is to calculate its binding energy. The mass of any atomic nucleus is less than the sum of the masses of the protons and neutrons which make it up. The difference between the two masses is called the binding energy, and it can be expressed in units of energy by using Einstein's well-known equation: $E = mc^2$, where m is the mass of a particle, E is the energy which corresponds to that mass (here, the binding energy), and c is the speed of light (3.00×10^8 m/sec). The greater the binding energy -- that is, the greater the amount of energy lost by the nucleus in the process of constructing it from protons and neutrons -- the more stable the nucleus.

Problem: Calculate the binding energy of a nucleus of carbon-14, given the following information[1]: Mass of ^{14}C Nucleus = 14.007682 amu. Mass of Proton = 1.008142 amu. Mass of Neutron = 1.008982 amu. 1 amu = 1.66×10^{-24} grams.

Solution: A $^{14}_{6}C$ nucleus contains 6 protons and 8 (14 - 6) neutrons. Therefore, the calculated mass of the carbon-14 nucleus is:

$$m_{calcd.} = (6 \times 1.008142 \text{ amu}) + (8 \times 1.008982 \text{ amu}) = 14.120708 \text{ amu.}$$

The observed mass of a carbon-14 nucleus is 14.007682 amu. Therefore:

Mass Difference = 14.120708 amu - 14.007682 amu = 0.113026 amu.

Mass Difference = (0.113026 amu)(1.66×10^{-24} g/amu) = 1.88×10^{-25} g.

Finally, convert this value into energy units, using Einstein's equation and the fact that 1 Joule = 1 kg m^2/sec^2. The result is:

Binding Energy = (1.88×10^{-25} g)(1 kg/1000 g)(3.00×10^8 m/sec)^2

Binding Energy = 1.69×10^{-11} kg m^2/sec^2 = 1.69 x 10^{-11} joules.

For comparison purposes, the binding energy per nucleon is used. (A "nucleon" is a proton or neutron.) In this case, (1.69×10^{-11} J)/14 = 1.21 x 10^{-12} J per nucleon.

---

1) Masses obtained from "The Atomic Nucleus" by R. D. Evans. 1955 by McGraw-Hill, Inc., page 137.

PARTICLE ACCELERATORS AND PARTICLE DETECTORS:

Transmutation of one element into another can be made to occur by bombarding an atomic nucleus with high-energy particles such as protons and electrons. For many years, the only way to obtain such high-energy particle beams was by using radioactive atoms as their source. It is now possible, however, to obtain beams of high-energy particles by using instruments known as particle accelerators. Particle accelerators use electric and magnetic fields to accelerate charged particles to high speeds. An example of a particle accelerator is the cyclotron, which uses magnetic fields to cause charged particles to move in circles. Linear particle accelerators are used if a straight-line pathway for the particles is desired -- the linear particle accelerator at Stanford University is about two miles long! Larger particle accelerators are projected for the future, among them the "superconducting super collider" which will be built underground in Texas.

The presence of high-energy particles and the pathways they travel can be detected using instruments known as particle detectors. Probably the best-known particle detector is the Geiger-Müller counter, which uses electronic methods to detect ionizing radiation. (Ionizing radiation is any form of radiation which strips electrons away from the atoms it encounters. Gamma radiation is one example of ionizing radiation.) The pathways traveled by particles can be detected using a cloud chamber, in which "trails" are left by particles which pass through the vapor from a volatile liquid. (This is similar to the "vapor trails" left by jet planes when they pass through the water vapor in the air.) Other particle detectors include the scintillation counter, which contains a substance such as zinc sulfide which glows when high-energy particles strike it (a "scintillation" is a spark or flash), and the dosimeter, which contains a piece of photographic film which gets darker when irradiated by gamma radiation. Nuclear chemists wear dosimeters daily.

MEASURING RADIOACTIVITY:

Radioactivity is measured in units called <u>Becquerels</u>. A radioactive sample which emits one particle per second has a radioactivity of one Becquerel. (The unit is named for the French scientist Henri Becquerel, who shared the 1903 Nobel Prize in Physics for his discovery of radioactivity.) A more convenient unit for most laboratory work is the <u>Curie</u>. One Curie = 3.7×10^{10} Becquerels, which is the radioactivity of 1.0 gram of radium. (The Curie is named in honor of Mme. Marie Curie, who shared the 1903 Nobel Prize in Physics with Becquerel and her husband Pierre. Mme. Curie also won the 1911 Nobel Prize in Chemistry for her discovery of the elements radium and polonium.)

Of greater interest are the units used to measure radiation <u>dosages</u> in living tissues. These are derived from a unit called the <u>Roentgen</u>, which is the amount of gamma radiation needed to ionize 2.08×10^9 air molecules per milliliter of air at STP. (The Roentgen is named for the German physicist Wilhelm Roentgen, who won the 1901 Nobel Prize in Physics for his discovery of <u>X-rays</u>, which were originally called "Roentgen Rays". X-rays consist of photons which generally have less energy than gamma rays. Their energy is measured in <u>electron volts (eV)</u>. One electron volt is the amount of energy gained by an electron which passes through a potential difference of one volt. 1.00 eV = 1.60×10^{-19} joules.) The most useful of the derived units for radiation dosage is the <u>rem</u>, which stands for "<u>R</u>oentgen <u>e</u>quivalent in <u>man</u>". One rem is the amount of radiation -- <u>any</u> kind, not just gamma radiation -- that produces the same biological effect in human tissues as would be produced by one roentgen of gamma radiation. The federal government has set a limit of 500 millirems per year as the maximum "safe" radiation dosage for the general public.[1] (Of course, the only <u>completely</u> safe dosage is <u>zero</u>, but this is unlikely.)

---

1) From "The Straight Dope" by Cecil Adams. 1984 by Chicago Reader, Incorporated, page 238.

EFFECTS OF RADIATION ON THE HUMAN BODY:

Each person on earth receives a certain dosage of radiation each year from natural sources. Much of this radiation is in the form of cosmic rays, which are largely made up of alpha, beta, and gamma particles which have their origin in nuclear reactions which are occurring in outer space. The gamma radiation is especially harmful because it is ionizing radiation -- it removes electrons from atoms, leaving positive ions behind. The problem is that if the lost electron was part of an electron pair, the resulting ion is a free radical. Free radicals have been shown to be probable cancer-causing agents. An average person in the United States receives about 100 millirems of "background radiation" each year from cosmic rays, X-rays, and other sources. (X-rays are produced by a process known as K-capture, in which an electron in the innermost shell of an atom (the "K" shell) is captured by the atomic nucleus, converting a proton into a neutron in the process: $_1^1p^+ + _{-1}^0e^- \longrightarrow _0^1n$. X-rays are used in medicine and dentistry to examine bones and teeth for structural irregularities, and in other ways as well.) The amount of "background radiation" varies with latitude, as the earth's magnetic field deflects a large amount of cosmic radiation away from the earth. People who travel in airplanes a great deal may receive more cosmic radiation than most, since they are not as shielded by the earth's magnetic field as those who remain on the earth's surface.[1] (The same is true for people who work in nuclear power plants!)

Radiation is used by the medical profession in certain beneficial ways. For example, iodine is absorbed by the thyroid gland, so radioactive iodine-131 can be given internally to people who have thyroid conditions as either a diagnostic or a therapeutic measure. Vitamin B_{12} contains cobalt ions, so a patient who has a Vitamin B_{12} deficiency may be fed some cobalt-60 so Vitamin B_{12} can be monitored.

---

1) From "The Straight Dope" by Cecil Adams. 1984 by Chicago Reader, Incorporated, page 238.

RADIOCARBON DATING:

Radiocarbon dating is a method used to determine the age of artifacts by measuring the amount of radioactive carbon-14 present in them. Carbon-14 is produced in the atmosphere when a free neutron collides with a nitrogen atom:

$$\ce{^{1}_{0}n + ^{14}_{7}N \longrightarrow ^{14}_{6}C + ^{1}_{1}H}.$$

The carbon-14 is oxidized to CO_2 by oxygen in the atmosphere, and the CO_2 is taken in by plants, which in turn are eaten by animals. Thus, every living thing keeps on ingesting radioactive carbon-14 until it dies. After the death of an organism, the carbon-14 present undergoes radioactive decay with a half-life of 5730 years. (Radioactive decay processes follow first-order kinetics, and therefore have half-lives which are constants. This makes determining the age of artifacts very easy.) By comparing the amount of carbon-14 present in an artifact with the amount present in living tissues, the age of the artifact can be calculated, as shown below.

Problem[1]: The radioactivity of the carbon-14 in a skull fragment is 0.016 Becquerels/gram. The radioactivity of carbon-14 in living tissue is 0.255 Becquerels/gram. Calculate the age of the skull fragment.

Solution: After each half-life, the radioactivity of the fragment will be half of what it was originally. Thus, after one half-life, the radioactivity is 0.255/2 = 0.128 Becquerels/gram; after two half-lives, it is 0.128/2 = 0.064 Bq/g; and so on. Therefore, the problem can be represented by the following equation:

$$(0.016 \text{ Becquerels/gram}) = (0.255 \text{ Becquerels/gram})(1/2)^n.$$

Solving this equation for n (either logarithmic methods or "trial and error" can be used) gives a value of 4, which means that four half-lives have elapsed since the death of the organism. Each half-life is 5730 years, so:

Age of Skull Fragment = 4 x 5730 years = 22,920 years.

---

1) From a similar problem on page 939 in "Chemistry: An Experimental Science" by G. M. Bodner and H. L. Pardue. 1989 by John Wiley & Sons, Inc.

NUCLEAR ENERGY:

Nuclear energy can be generated by either of two processes: nuclear fission or nuclear fusion. Nuclear <u>fission</u> involves the splitting of one large atomic nucleus into smaller atomic nuclei. Consider the following nuclear reaction:

$$^{235}_{92}U + ^{1}_{0}n \longrightarrow ^{94}_{36}Kr + ^{139}_{56}Ba + 3\,^{1}_{0}n + energy.$$

The above equation shows that bombarding a uranium-235 nucleus with a neutron gives two smaller nuclei and three neutrons, in addition to giving off large amounts of energy. The three neutrons produced can each strike another uranium nucleus and cause the reaction to occur again, unless they are absorbed by some other substance such as cadmium. This is the basis for a nuclear "chain reaction", and is the basic process used in nuclear power plants. "Meltdowns" occur when too much energy or too many neutrons are produced in a short period of time. Nuclear reactors are now being designed in better ways so as to make meltdowns less likely.[1]

Nuclear <u>fusion</u> involves the joining of two or more small atomic nuclei to form one large atomic nucleus. Nuclear fusion occurs in stars, which give off large amounts of energy in the process of fusing hydrogen nuclei to form helium:

$$^{2}_{1}H + ^{3}_{1}H \longrightarrow ^{4}_{2}He + ^{1}_{0}n + energy.$$

(In a sense, then, solar energy is nuclear energy, since sunlight is produced by nuclear processes.) The above reaction produces 17.0×10^{6} electron volts of energy per helium atom produced, so fusion research is being actively pursued as a possible source of energy. (In March of 1989, scientists at the University of Utah announced that they had achieved nuclear fusion in a beaker of water, but subsequent investigations showed their claims to be largely without merit.[2]) The raw material -- hydrogen -- for the fusion process is readily available from sea water (H_2O), but it is extremely difficult to get two positively-charged nuclei close enough to fuse.

---

1) <u>U. S. News & World Report</u>, May 29, 1989, pages 52-53.
2) For a full report, see <u>Newsweek</u>, May 8, 1989, pages 48-56.

ORGANIC CHEMISTRY:

Organic chemistry is best defined as the chemistry of the compounds that make up living organisms -- hence the name, "organic" chemistry. Organic chemistry is also defined as the chemistry of carbon, since most of the interesting compounds which make up living things contain carbon. Carbon atoms have a strong tendency to form covalent bonds to each other, forming long strands or "chains" of atoms which are the structural basis for many biologically important molecules. Millions of different organic compounds are known to exist, and more are discovered each day. In addition, synthetic organic materials such as Nylon and Teflon, which are not found in nature, have made significant impacts on the ways people live.

In addition to carbon, the elements usually found in organic compounds are hydrogen, oxygen, nitrogen, and the halogens, with sulfur, phosphorus, and some others occasionally appearing. Since all of these elements are non-metals, they tend to form _covalent_ bonds to each other. The _number_ of covalent bonds attached to an atom of a non-metal can be predicted using the equation: $B = 8 - GN$, where "GN" is the group number of the group in the Periodic Table in which the element is located. Thus, carbon atoms have _four_ covalent bonds ($8 - 4 = 4$), nitrogen atoms have _three_, oxygen atoms have _two_, and halogen atoms have _one_ covalent bond. Atoms of hydrogen are the major exception to this rule, having _one_ covalent bond each.

The geometry of organic molecules is determined by applying the VSEPR theory to each atom in the molecule. This implies that atoms in organic molecules use hybrid orbitals to form bonds. Since most of the elements mentioned above are second-row elements, d orbitals do not usually enter the bonding picture. The usual hybridizations of atoms in organic molecules are sp^3 (tetrahedral geometry), sp^2 (planar triangular geometry), and sp (linear geometry). It is important to keep this in mind when viewing two-dimensional drawings of three-dimensional molecules.

The simplest class of organic compounds are the hydrocarbons, which are composed of only hydrogen and carbon. The simplest hydrocarbons are the alkanes, which have the general formula C_nH_{2n+2}. The simplest alkane is methane, whose molecules contain one carbon atom each. Since $(2)(1) + 2 = \underline{4}$, the formula for methane is CH_4. Using the basic rules for drawing Lewis Structures, we would draw a molecule of methane with the carbon atom in the center, surrounded by four atoms of hydrogen. Several different ways to draw a molecule of methane are shown below.

The first two drawings are simple Lewis Structures, with dashes representing bonds or dots representing electrons. The third drawing is an attempt to represent a three-dimensional molecule on a two-dimensional sheet of paper. The wedge-shaped line represents a bond coming out of the plane of the page toward the reader; the dotted line represents a bond going behing the plane of the page away from the reader. (The solid lines represent bonds lying in the plane of the page.) This notation is sometimes used tö emphasize the three-dimensional nature of molecules.

The next simplest alkane, ethane, has the formula C_2H_6. Several ways to represent molecules of ethane are shown below.

The first two drawings are simple Lewis structures. The other two drawings are called condensed structures. The carbon-carbon bond in ethane is obvious in the Lewis structures, but not as obvious in the condensed structures. Practice looking at condensed structures to see the carbon-carbon bonds. For example, propane (C_3H_8) contains two carbon-carbon bonds. Find them: $CH_3-CH_2-CH_3$ $CH_3CH_2CH_3$

ISOMERS OF ALKANES:

Methane (CH_4), ethane (C_2H_6), and propane (C_3H_8) can each be uniquely identified just by their molecular formulas. Drawing the structure for butane (C_4H_{10}) poses a problem, however, since there are two known compounds which have the formula C_4H_{10}. Different compounds which have the same molecular formula are known as isomers. The two isomers of butane are known as "n-butane" (the "n" stands for "normal") and "isobutane" (an isomer of butane). Their structures are shown below.

n-butane: $CH_3-CH_2-CH_2-CH_3$ $CH_3CH_2CH_2CH_3$ $CH_3(CH_2)_2CH_3$

isobutane: $CH_3-CH-CH_3$ CH_3CHCH_3 $CH_3CH(CH_3)_2$
                    $|$ $|$
                    CH_3 CH_3

The different structures of the molecules of n-butane and isobutane give rise to the different properties of the compounds. For example, n-butane has a boiling point of 0 ^{O}C, while isobutane has a boiling point of -12 ^{O}C. (The more elongated n-butane molecule has a greater surface area than the isobutane molecule, thus increasing the London forces of attraction between n-butane molecules and increasing the boiling point.) Thus, the compounds are different, but have the same formula.

The number of possible isomers increases dramatically as the number of carbon atoms in the formula of the compound increases. For pentane (C_5H_{12}), three isomers are possible -- n-pentane, isopentane, and neopentane ("neo" means "new"):

n-pentane: $CH_3CH_2CH_2CH_2CH_3$

isopentane: $CH_3CH_2CH(CH_3)_2$

neopentane:
$$CH_3-\overset{\displaystyle CH_3}{\underset{\displaystyle CH_3}{\overset{|}{\underset{|}{C}}}}-CH_3 \;=\; CH_3C(CH_3)_3$$

For hexane (C_6H_{14}), five isomers exist. For heptane (C_7H_{14}), nine isomers exist. Larger alkanes have even greater numbers of isomers. Inventing new prefixes for the name of each isomer of the larger alkanes is clumsy, tedious, and inefficient -- a better system of nomenclature for alkanes is needed. The present system for naming the alkanes was developed by the International Union of Pure and Applied Chemistry.

NOMENCLATURE OF ALKANES:

The IUPAC system of nomenclature for the alkanes is based on three very simple rules, which are listed below.

1. The <u>longest</u> "chain" of carbon atoms present in the molecule is used to determine the "base" for the name of the alkane. The "base" names are <u>methane</u> (C_1), <u>ethane</u> (C_2), <u>propane</u> (C_3), <u>butane</u> (C_4), <u>pentane</u> (C_5), <u>hexane</u> (C_6), <u>heptane</u> (C_7), <u>octane</u> (C_8), <u>nonane</u> (C_9), <u>decane</u> (C_{10}), and so on. (Notice that beginning with pentane, Greek prefixes are used -- "pent-" means "5", "hex-" means "6", etc.)

2. "Side chains" (substituents attached to the "base chain") are named for the number of carbon atoms they contain, using the above names with the "-ane" ending replaced by "<u>-yl</u>". Thus, a CH_3- substituent is called a <u>methyl</u> group, a CH_3CH_2- "side chain" is called an <u>ethyl</u> group, $CH_3CH_2CH_2$- is a <u>propyl</u> group, etc.

3. The location of the "side chains" on the "main chain" is indicated by numbering the main chain and placing the number before the name of the side chain, separated from it by a hyphen. Since the "main chain" can be numbered from either end, the numbers should be chosen so that the <u>smallest numbers possible</u> are the numbers used in the name of the compound.

<u>Problem</u>: Name the following compound: $CH_3-CH_2-\overset{\overset{\displaystyle CH_3}{|}}{CH}-CH_3$.

<u>Solution</u>: The "main chain" is four carbon atoms long, so the base name is <u>butane</u>. There is one 1-carbon "side chain", which is called a <u>methyl</u> group. If the four-carbon "chain" is numbered 1 through 4 from left to right, carbon atom # 3 has the methyl group, but if the numbering is done from right to left, carbon atom # 2 has the methyl group. The smaller number is preferred -- <u>2-methylbutane</u>.

<u>Problem</u>: Name the following compound: $CH_3-CH_2-CH_2-CH_2-\overset{\overset{\displaystyle CH_3}{|}}{CH}-CH_2-\overset{\overset{\displaystyle CH-CH_2CH_3}{|}}{C}-CH_2CH_3$
$\overset{|}{CH_3}$

<u>Solution</u>: <u>3,6-dimethyl-4-ethyldecane</u>.

The <u>longest</u> (bent!) "chain" = C_{10}. "di-" = <u>two</u>. (Two CH_3-'s present.)

PROPERTIES AND REACTIONS OF ALKANES AND CYCLOALKANES:

Cycloalkanes are similar to alkanes in the sense that they contain only carbon and hydrogen atoms, and that all of the atoms present are connected to each other by single covalent bonds. Where cycloalkanes differ from other alkanes is in their structures and formulas. Cycloalkanes have their carbon atoms arranged in a cyclic fashion, resulting in a general formula of C_nH_{2n} for cycloalkanes. The prefix "cyclo-" is used before the "base" name to indicate a cycloalkane. The structures of cyclopropane, cyclobutane, and cyclohexane are given below.

Frequently, molecules like these are simply represented by the geometric forms they resemble -- cyclopropane is △ , cyclobutane is ▢ , and cyclohexane is ⬡ . Since each line in the above figures represents a carbon-carbon covalent bond, this way of drawing organic molecules is sometimes called bond-line notation.

Alkanes and cycloalkanes are relatively unreactive as organic compounds go. However, they do undergo two reactions readily. One of these is oxidation, or combustion. This should be no surprise to anyone who has ever used a propane stove, a butane cigarette lighter, or high-octane gasoline for automobiles. The other reaction is halogenation -- the substitution of a halogen atom for a hydrogen atom in an alkane or cycloalkane. The reaction occurs by a free-radical process, in which halogen radicals remove hydrogen atoms from the alkane to form alkyl radicals, which then react with halogen molecules to form the alkyl halides. (See below.)

$$CH_4 + Cl\cdot \longrightarrow CH_3\cdot + HCl$$
$$CH_3\cdot + Cl_2 \longrightarrow CH_3Cl + Cl\cdot$$

(These are the propagation steps in the free-radical chain reaction mechanism.)

Methane, ethane, propane, and butane are gases at room temperature. Most other alkanes are oily liquids or waxy solids which are less dense than water.

STRUCTURE, GEOMETRY, AND NOMENCLATURE OF ALKENES:

The molecules of alkanes and cycloalkanes contain only sp^3-hybridized carbon atoms, since all of the carbon-carbon bonds are <u>single</u> bonds. <u>Alkenes</u> are hydrocarbons which contain at least one carbon-carbon <u>double</u> bond. Therefore, at least two carbon atoms in the molecules of alkenes must be sp^2-hybridized. (Note the spellings, by the way: alk<u>a</u>ne vs. alk<u>e</u>ne. Be careful!) The general formula for an alkene is C_nH_{2n} -- the same as the general formula for cycloalkanes. Since alkenes and cycloalkanes contain <u>less</u> than the maximum number of hydrogen atoms possible per molecule, they are said to be <u>unsaturated</u>, whereas alkanes are said to be <u>saturated</u>. ("Polyunsaturated" fats and oils contain many double bonds.)

The simplest alkene is <u>ethene</u> (or <u>ethylene</u>), which has the formula C_2H_4. The structure of ethylene appears at the right. Note the geometry of the molecule -- each carbon atom is sp^2-hybridized, so the bond angles should be equal to approximately 120^0. The condensed formula for ethene is $CH_2=CH_2$. The nomenclature of alkenes is based on the nomenclature of alkanes, except the "-ane" ending is replaced by "-ene". For example, $CH_2=CH-CH_3$ is <u>propene</u> (sometimes called <u>propylene</u>). One additional problem arises with the larger alkenes, however: the location of the double bond must be specified. Again, the problem is solved by numbering the longest "chain" of carbon atoms which contains the double bond, and writing the number of the lower-numbered carbon atom in front of the name of the alkene, separated from it by a hyphen. For example, $CH_2=CH-CH_2-CH_3$ is <u>1-butene</u>, whereas $CH_3-CH=CH-CH_3$ is <u>2-butene</u>. (These two compounds are <u>isomers</u> of C_4H_8.)

Problem: Name the compound shown at the right:

Solution: The longest "chain" which includes <u>both</u> of the sp^2-hybridized carbon atoms is <u>six</u> carbon atoms long. Thus, the "base" name is "hexene". The full name is <u>2-methyl-3-propyl-2-hexene</u>.

STEREOISOMERS AND REACTIONS OF ALKENES:

The <u>pi</u> bond in a carbon-carbon double bond is formed by the "sideways" overlap of two <u>p</u> orbitals. Since the carbon atoms and their substituents must be aligned in a particular configuration for this to happen, there is a certain amount of rigidity in alkene molecules that is not present in alkanes. This leads to some interesting results. Consider, for example, the two drawings of 2-butene below:

$$CH_3\text{-}C\text{=}C\text{-}CH_3 \text{ (both H below)} \quad (m.p. -139\ ^oC) \qquad CH_3\text{-}C\text{=}C\text{-}H / CH_3 \quad (m.p. -105\ ^oC)$$

These are obviously different compounds, as shown by the fact that their melting points differ by 34 ^oC. Yet both compounds are 2-butene! These are isomers which differ only in the arrangement of their atoms in three-dimensional space, and are called <u>stereoisomers</u>. The compound on the left is called <u>cis</u>-2-butene, and the compound on the right is called <u>trans</u>-2-butene. (To remember this, notice that the carbon atoms in the <u>cis</u> isomer are arranged in the shape of a "C". "C" for "cis"!) Isomers which differ only by being <u>cis</u> or <u>trans</u> are called <u>geometric isomers</u>.

Alkenes typically undergo <u>addition</u> reactions, as opposed to the typical <u>substitution</u> reactions which alkanes undergo. In an addition reaction, the pi bond of the carbon-carbon double bond is broken, and single bonds are formed to each of two other atoms. These reactions are called addition reactions because the formula of the product can be found by simply adding the formulas of the reactants. Some examples of addition reactions are shown below, with <u>propene</u> (C_3H_6) as the alkene.

$CH_2\text{=}CH\text{-}CH_3 + H_2 \xrightarrow{Pt} CH_3\text{-}CH_2\text{-}CH_3.$ (product = C_3H_8.)

$CH_2\text{=}CH\text{-}CH_3 + Br_2 \longrightarrow Br\text{-}CH_2\text{-}CHBr\text{-}CH_3.$ (product = $C_3H_6Br_2$.)

$CH_2\text{=}CH\text{-}CH_3 + HCl \longrightarrow CH_3\text{-}CHCl\text{-}CH_3.$ (product = C_3H_7Cl.)

$CH_2\text{=}CH\text{-}CH_3 + H_2O \xrightarrow{H^+} CH_3\text{-}CHOH\text{-}CH_3.$ (product = C_3H_8O.)

$CH_2\text{=}CH\text{-}CH_3 + O_3 \xrightarrow{Zn} CH_2\text{=}O + O\text{=}CH\text{-}CH_3.$ (You can't win 'em all!)

ALKYNES:

Compounds which contain at least one carbon-carbon triple bond are called alkynes. (Again, be careful of the spelling -- alk<u>y</u>nes, as opposed to alk<u>a</u>nes or alk<u>e</u>nes.) Since the carbon atoms in carbon-carbon triple bonds are <u>sp</u>-hybridized, the presence of the triple bond imparts a <u>linear</u> geometry to the alkyne molecule. Thus, no "cis-trans" isomerism is possible for alkynes.

The nomenclature for alkynes is similar to the nomenclature for alkenes. The simplest alkyne, C_2H_2, is called <u>ethyne</u>. (It is commonly called <u>acetylene</u>, however, which is confusing, since it isn't an alkene!) The structure of ethyne is $H-C\equiv C-H$, and the geometry of the molecule is the linear geometry shown at the left. Similarly, <u>propyne</u> is $H-C\equiv C-CH_3$, and <u>2-butyne</u> is $CH_3-C\equiv C-CH_3$. These examples show that the generic formula for an alkyne is $\underline{C_nH_{2n-2}}$ -- for example, propyne is C_3H_4.

The carbon-carbon triple bonds present in alkynes are composed of one sigma bond and <u>two</u> pi bonds. Therefore, it shouldn't be too surprising to observe that alkynes undergo the same kinds of <u>addition</u> reactions that alkenes do, except for the fact that the alkynes can react with <u>twice</u> as much of a particular reactant as the alkenes can. For example, propyne reacts with bromine in a 1:2 molar ratio:

$$H-C\equiv C-CH_3 \ + \ 2 \ Br_2 \ \longrightarrow \ Br_2CH-CBr_2-CH_3. \quad (\text{product} = C_3H_4Br_4.)$$

Another reaction which some alkynes can undergo is an acid-base reaction. Alkynes which have a hydrogen atom attached to an sp-hybridized carbon atom are more acidic than most other hydrocarbons, and strong bases (such as $NaNH_2$) can react with them:

$$H-C\equiv C-CH_3 \ + \ NaNH_2 \ \longrightarrow \ NaC\equiv C-CH_3 \ + \ NH_3.$$

This is an important reaction in the synthesis of large organic molecules from small organic molecules, because the negatively-charged carbon atom in $NaC\equiv C-CH_3$ reacts with positively-charged carbon atoms in other molecules, forming new bonds as shown:

$$NaC\equiv C-CH_3 \ + \ CH_3-CH_2-Br \ \longrightarrow \ CH_3-CH_2-C\equiv C-CH_3 \ + \ NaBr.$$

AROMATIC HYDROCARBONS:

The word "aromatic" is not a reference to the way hydrocarbons smell! <u>Aromatic hydrocarbons</u> are a class of hydrocarbons which are unusually stable due to the presence of certain structural features in their molecules. Aromatic molecules generally contain <u>cyclic</u> arrangements of <u>sp^2</u>-hybridized carbon atoms. The carbon atoms all lie in the same plane, so that the "leftover" p orbitals can all overlap with each other. As a result, the electrons in these p orbitals are free to move all over the ring of carbon atoms. This extensive delocalization of pi electrons is the structural feature that gives aromatic hydrocarbons their unusual stability.

A good example of an aromatic hydrocarbon is <u>benzene</u> (C_6H_6). The six carbon atoms are arranged in a ring, with all bond angles = 120°. Each carbon atom is connected to one hydrogen atom. "Bond-line" structures for benzene appear below:

The two structures at the left are <u>resonance</u> structures. The dotted lines in the third structure represent the delocalized electrons. Benzene is usually drawn as shown in the structure at the right -- the circle represents the dotted lines.

The presence of carbon-carbon double bonds in the resonance structures might suggest that benzene would undergo addition reactions in a way similar to the way alkenes do. However, additions of this type would destroy the aromatic nature of the benzene molecule. Since the aromaticity of benzene is a stabilizing factor, reactions of this type are unlikely. Rather, benzene and other aromatic compounds tend to undergo <u>substitution</u> reactions, similar to those which alkanes undergo. In an aromatic substitution reaction, a hydrogen atom is removed from the benzene ring and replaced with some other atom or group of atoms. An example is given below:

$$C_6H_6 + Br_2 \xrightarrow{Fe} C_6H_5Br + HBr. \text{ (The iron (Fe) is a catalyst.)}$$

FUNCTIONAL GROUPS:

Hydrocarbons are relatively unreactive organic compounds. Most organic compounds contain electronegative elements such as oxygen, nitrogen, or halogens. The presence of electronegative atoms in organic molecules creates <u>polar</u> regions within these molecules. Much of the chemistry of organic molecules is attributed to the attraction of a positive electrical charge on an atom for a partial negative charge on another atom. The polar regions within organic molecules are known as <u>functional groups</u>, since their presence helps to determine how an organic molecule will function in a particular reaction. Organic compounds are often classified according to their functional groups. The table below lists some of the functional groups commonly found in organic molecules and gives examples of each.

Functional Group	Generic Formula*	Specific Example
Alcohol	R-OH	CH_3-CH_2-OH
Ether	R-O-R'	$CH_3-CH_2-O-CH_2-CH_3$
Amine	R-NH$_2$	$C_6H_5-NH_2$
Carboxylic Acid	R-CO-OH	CH_3-CO_2H
Ester	R-CO-OR'	$CH_3-CO-O-CH_2-CH_3$
Amide	R-CO-NH$_2$	$CH_3-CH_2-CH_2-CO-NH_2$
Aldehyde	R-CO-H	C_6H_5-CHO
Ketone	R-CO-R'	$C_6H_5-CO-CH_3$
Nitrile	R-C≡N	$N≡C-CH_2-CH_2-CH_2-CH_2-C≡N$

*In the above table, the symbols R and R' are used to represent the hydrocarbon part of the molecule. R and R' can be methyl groups, ethyl groups, or any other hydrocarbon unit. In any one molecule, R and R' may or may not be equal. Also, it should be understood that the -CO- fragment found in several of the above functional groups is a carbon-oxygen <u>double</u> bond (C=O), also called a <u>carbonyl group</u>.

ALCOHOLS, ETHERS, AND AMINES:

Alcohols are compounds which contain one or more hydroxy groups (-OH) attached to sp^3-hybridized carbon atoms. Among the better-known alcohols are ethyl alcohol (CH_3-CH_2-OH), which is used in "alcoholic" beverages, and methanol (CH_3-OH), which is used in some automobiles for fuel. Alcohols can be synthesized in many ways, among them the acid-catalyzed addition of water to an alkene:

$$CH_3-CH=CH-CH_3 + H_2O \xrightarrow{H^+} CH_3-CHOH-CH_2-CH_3. \quad \text{(Product = 2-butanol.)}$$

Alcohols can be easily oxidized, as in the intoxicating reaction of ethyl alcohol:

$$CH_3-CH_2-OH + O_2 \longrightarrow CH_3-CO_2H + H_2O.$$

Alcohols can also undergo elimination (the reverse of addition) to form alkenes:

$$CH_3-CH_2-CH_2-OH + H_3O^+ \longrightarrow CH_3-CH=CH_2 + H_2O + H_3O^+.$$

Ethers are compounds which contain one or more oxygen atoms attached to two different carbon atoms. A common ether is diethyl ether ($CH_3-CH_2-O-CH_2-CH_3$), which is used in biology laboratories as an anaesthetic and in chemical research as a solvent. Ethers can be formed by the acid-catalyzed reaction of alcohols:

$$2\ CH_3-CH_2-OH + H_2SO_4 \longrightarrow CH_3-CH_2-O-CH_2-CH_3 + H_2O + H_2SO_4.$$

Strong acids such as HI can cleave ethers, forming alcohols and alkyl halides:

$$CH_3-CH_2-O-CH_2-CH_3 + HI \longrightarrow CH_3-CH_2-OH + CH_3-CH_2-I.$$

Amines are essentially molecules of ammonia (NH_3) in which one or more hydrogen atoms have been replaced by hydrocarbon groups. (Hydrocarbon groups which are derived from alkanes are usually called alkyl groups.) Like ammonia, amines are weak Brønsted bases, with K_b values equal to approximately 1×10^{-4}:

$$CH_3-NH_2 + H_2O \rightleftharpoons CH_3-NH_3^+ + OH^-.$$

Like ammonia, amines have a pungent odor similar to the smell of fish. Aromatic amines such as aniline ($C_6H_5-NH_2$) are used in the production of azo dyes by the textiles industry. (Azo compounds contain a nitrogen-nitrogen double bond.)

CARBOXYLIC ACIDS, ESTERS, AND AMIDES:

Carboxylic acids are compounds which contain one or more hydroxy groups (-OH) directly attached to carbonyl groups (C=O). The resulting $-CO_2H$ group reacts by donating the proton, thus behaving as a Brønsted acid. Carboxylic acids are weak acids, with K_a values typically about 1×10^{-5}:

$$CH_3-CO_2H + H_2O \rightleftharpoons CH_3-CO_2^- + H_3O^+.$$

Two of the better-known carboxylic acids are acetic acid (CH_3-CO_2H), the acid found in vinegar, and formic acid ($H-CO_2H$), the acid which causes the burning sensation from the bite of a red ant. Carboxylic acids can be made by oxidizing alcohols:

$$CH_3-CH_2-OH + O_2 \longrightarrow CH_3-CO_2H + H_2O.$$

Esters are essentially carboxylic acids in which the proton in the $-CO_2H$ group has been replaced by an alkyl group. Esters are generally known for their pleasant odors. Examples include pentyl acetate ($CH_3CH_2CH_2CH_2CH_2-O-CO-CH_3$), known as "pear oil" for its fruitlike aroma, and methyl salicylate ($HOC_6H_4-CO_2CH_3$), sometimes called "oil of wintergreen" for its wintergreen smell. Esters can be made by the acid-catalyzed reaction of a carboxylic acid with an alcohol:

$$CH_3-CO_2H + HO-CH_2CH_2CH_2CH_3 \longrightarrow CH_3-CO-O-CH_2CH_2CH_2CH_3 + H_2O.$$

When this reaction is carried out using a dicarboxylic acid (two $-CO_2H$ groups in one molecule) and a diol (short for "di-alcohol"; two -OH groups in one molecule), the product is a polymer known as a "poly-ester"! For example, Dacron is the product of the reaction between $HO-CH_2-CH_2-OH$ and $HO_2C-C_6H_4-CO_2H$.

Amides are essentially carboxylic acids in which the -OH group in the $-CO_2H$ group has been replaced by an amino group ($-NH_2$). Amides can be formed by the acid-catalyzed reaction of carboxylic acids with ammonia or amines:

$$C_6H_5-CO_2H + NH_3 \longrightarrow C_6H_5-CO-NH_2 + H_2O.$$

Polyamides (such as Nylon) are formed by the reaction of diacids with diamines.

ALDEHYDES, KETONES, AND NITRILES:

Aldehydes are compounds which contain one or more carbonyl groups (C=O) connected to at least one hydrogen atom. Some of the better-known aldehydes are formaldehyde ($H_2C=O$), which is used to preserve biological specimens, and benzaldehyde ($C_6H_5-CH=O$), which has an almond-like odor and is used as a food additive. Aldehydes can be synthesized by the partial oxidation of alcohols:

$$2 CH_3OH + O_2 \longrightarrow 2 H_2C=O + 2 H_2O.$$

(The above reaction shows why methanol (CH_3OH) should not be consumed internally -- it reacts with oxygen in the bloodstream to form formaldehyde!) Aldehydes can be further oxidized to form carboxylic acids, as shown for benzaldehyde below:

$$2 C_6H_5-CH=O + O_2 \longrightarrow 2 C_6H_5-CO_2H.$$

Ketones are compounds which contain one or more carbonyl groups which are connected to two different carbon atoms. One of the most commonly used ketones is acetone ($CH_3-CO-CH_3$), which is the major component of fingernail polish remover. The chemistry of ketones is very similar to that of aldehydes -- ketones can also be formed by the oxidation of alcohols, for example. Ketones and aldehydes can both be reduced to form alcohols, as shown below for the reduction of acetone:

$$(CH_3)_2C=O + H_2 \xrightarrow{Pd} (CH_3)_2CH-OH. \quad \text{(Palladium is a catalyst.)}$$

(The product of the above reaction is isopropyl alcohol, or "rubbing alcohol".)

Nitriles are compounds which contain one or more carbon-nitrogen triple bonds (C≡N). A common nitrile is adiponitrile ($N≡C-CH_2-CH_2-CH_2-CH_2-C≡N$), which is used as a raw material for the production of Nylon. Nitriles can be prepared by the reaction of alkyl halides with cyanide ions (CN^-), as shown below:

$$Br-CH_2-CH_2-CH_2-CH_2-Br + 2 CN^- \longrightarrow N≡C-CH_2-CH_2-CH_2-CH_2-C≡N + 2 Br^-.$$

Nitriles are very versatile compounds -- they can be reduced to form amines, oxidized to form amides, or hydrolyzed to form carboxylic acids.

STEROIDS, FATS, AND SOAPS:

The presence or absence of polar functional groups in the molecules of an organic compound affect the behavior of the compound in living tissues. For example, the class of compounds known as steroids are mostly carbon and hydrogen, and are therefore insoluble in water (hydrophobic) but soluble in fatty tissues (lipophilic).

The basic hydrocarbon skeletal structure of steroid molecules is shown above. Some well-known steroids include cholesterol (an alcohol, since its name ends in "-ol") and the sex hormones testosterone (a ketone) and estradiol (two -OH groups present). The accumulation of cholesterol in fatty tissues poses serious health problems.

Fats are largely composed of esters made from the alcohol glycerol ($HO-CH_2-CHOH-CH_2-OH$) and three fatty acids. Fatty acids are carboxylic acids in which the carbonyl group is attached to a long hydrocarbon chain (usually eight to eighteen carbon atoms long). Saturated fats are those in which only carbon-carbon single bonds are present in the hydrocarbon chains; unsaturated fats contain carbon-carbon double bonds. (If more than one C=C bond is present in the fat molecule, it is a polyunsaturated fat.) Because of these long hydrocarbon chains, fat molecules are largely insoluble in water. However, treatment of fats (or any other esters) with sodium hydroxide (NaOH, "lye") cleaves the ester linkage, resulting in the formation of a salt which has an ionic, "polar" end and a non-ionic, "nonpolar" end:

$$R-O-CO-R' + OH^- \longrightarrow R-O-H + {}^-O-CO-R'.$$

Salts like these are called soaps, and the reaction above is called saponification (from the Latin for "soap-making"). Soaps work by dissolving grease or fat in the hydrocarbon parts of soap molecules. This surrounds the grease globule with a cluster of soap molecules known as a micelle, in which the polar, ionic ends are outermost. Since these are water-soluble, the grease globule can be washed away.

Polymers (sometimes called macromolecules) are large molecules made by joining thousands of smaller molecules. The molecular weights of polymers may thus be about 1×10^6 amu or more. Polymers are classified as either addition polymers or condensation polymers. The formula of an addition polymer is just the sum of the formulas of the smaller molecules (called monomers, since "mono" means "one" and "poly" means "many") from which it is made. An example of an addition polymer is polyethylene, which is made by connecting a large number of ethylene molecules:

$$n\ CH_2=CH_2 \longrightarrow \ldots-CH_2-CH_2-CH_2-CH_2-CH_2-CH_2-\ldots = (CH_2-CH_2)_n.$$

In forming a typical condensation polymer, molecules of two different compounds are allowed to react, and the formula of the resulting polymer is not the sum of the formulas of the reacting monomers. An example of a condensation polymer is Nylon, which is made by the reaction of a dicarboxylic acid (or a more reactive derivative thereof) with a diamine. Water is formed as a by-product of this process:

$$n\ HO-CO-(CH_2)_4-CO-OH + n\ H_2N-(CH_2)_6-NH_2 \longrightarrow$$

$$(CO-(CH_2)_4-CO-NH-(CH_2)_6-NH)_n + n\ H_2O.$$

Polyethylene is commonly used in many plastics, and Nylon is used to make stockings and ropes. Other polymers and their uses are listed in the table below.

Monomer(s)	Polymer	Name	Use
Addition Polymers:			
$CF_2=CF_2$	$(CF_2-CF_2)_n$ (polytetrafluoroethylene)	Teflon	Non-stick coatings
$CH_2=CH-C{\equiv}N$ (acrylonitrile)	$(CH_2-CH)_n$ CN (polyacrylonitrile)	Orlon	Acrylic fibers
Condensation Polymer:			
$HO-CH_2-CH_2-OH$, $HO-CO-C_6H_4-CO-OH$	$(CO-C_6H_4-CO-O-CH_2-CH_2-O)_n$ (poly(ethylene terephthalate))	Dacron	Textile fibers

AMINO ACIDS, PROTEINS, AND ENZYMES:

Amino acids are carboxylic acids which contain an amino group ($-NH_2$). Most of the twenty amino acids which are essential in human nutrition have an amino group attached to the carbon atom which is next to the carbonyl group. Thus, the "generic" formula for an amino acid is $H_2N-CHR-CO_2H$, where R is usually an alkyl group. Examples of amino acids include glycine ($H_2N-CH_2-CO_2H$, R = H), alanine ($H_2N-CH(CH_3)-CO_2H$, R = CH_3), and serine ($H_2N-CH(CH_2OH)-CO_2H$, R = CH_2OH). Since the $-CO_2H$ group is acidic and the $-NH_2$ group is basic, a proton (H^+) from the $-CO_2H$ group is usually transferred to the $-NH_2$ group, resulting in $H_3\overset{+}{N}-CHR-CO-O^-$ as the most stable form of most amino acids.

An interesting feature of the geometry of amino acids is illustrated in the two drawings of alanine shown at the right. The drawings are mirror images of each other, but the molecules to which they correspond are not the same -- they cannot be superimposed on each other. (A good analogy is that the molecules above are related in the same way in which a left hand and a right hand are related.) These two molecules are stereoisomers -- they differ only in the three-dimensional arrangement of their atoms. Stereoisomers which are "mirror images" are called enantiomers. Molecules which are not the same as their "mirror images" are called chiral (from the Greek "cheir", meaning "hand"). Amino acids can be either right-handed or left-handed!

When amino acids react with each other, the $-CO_2H$ group of one amino acid and the $-NH_2$ group of another combine to form an amide linkage ($-CO-NH-$), which is sometimes called a peptide bond. Polypeptides are long strands of amino acids which are connected by peptide bonds. Proteins and enzymes are clusters of one or more polypeptides which perform specific biological functions. Enzymes react with specific chiral molecules like a left-handed glove specifically fits a left hand.

CARBOHYDRATES AND NUCLEIC ACIDS:

Carbohydrates are organic compounds with the general formula $C_n(H_2O)_m$ -- that is, they are "hydrates of carbon", hence the name "carbo-hydrates". They are functionally aldehydes and ketones which also contain one or more hydroxy (-OH) groups. Examples of simple carbohydrates in which n = m (called monosaccharides) include glucose ($C_6H_{12}O_6$, O=CH-CHOH-CHOH-CHOH-CHOH-CH$_2$OH) and fructose ($C_6H_{12}O_6$, HOCH$_2$-CO-CHOH-CHOH-CHOH-CH$_2$OH). These compounds are simple sugars, which is where the term "monosaccharide" comes from. Glucose is an example of an aldose (a simple sugar which contains an aldehyde functional group), and fructose is an example of a ketose (a simple sugar which contains a ketone functional group.

Carbohydrates in which n ≠ m are called polysaccharides. A common example is ordinary table sugar, sucrose ($C_{12}H_{22}O_{11}$). A molecule of sucrose is made from a molecule of glucose and a molecule of fructose, with a water molecule as the by-product of the reaction. Sucrose is thus a disaccharide -- two molecules of monosaccharides combine to form one molecule of the disaccharide. The molecules of the monosaccharides are present in polysaccharides in a cyclic form known as a hemiacetal or hemiketal (for aldoses and ketoses, respectively). Other examples of polysaccharides include amylose (in starch) and cellulose (in plant cell walls).

Nucleic acids are the polymers which carry genetic information in human cells. They are made of monosaccharides, cyclic amines (called "nitrogen bases"), and polyphosphate units. They are commonly known as either RNA (ribonucleic acid, derived from the monosaccharide ribose) or DNA (deoxyribonucleic acid, derived from the monosaccharide deoxyribose). The well-known "double helix" structure of DNA was discovered by the American scientist James Watson and the British scientists Francis Crick and Maurice Wilkins, who shared the 1962 Nobel Prize in Physiology/ Medicine for this work. Hydrogen bonds connect the two strands of the double helix.

118

APPENDIX -- THE LAWS OF LOGARITHMS:

The <u>logarithm</u> of a number is the power to which <u>ten</u> (10) must be raised to obtain the number. The <u>natural logarithm</u> of a number is the power to which <u>e</u> (e = 2.71828...) must be raised to obtain the number. The abbreviation for "logarithm" is "log"; the abbreviation for "natural logarithm" is "ln". Some of the laws of logarithms with which you should be familiar are listed below.

Law of Logarithms	Example
$\log 10^a = a$	$\log 10{,}000 = \log 10^4 = 4.$
$10^{(\log a)} = a$	$10^{(\log 10{,}000)} = 10^4 = 10{,}000.$
$\log a^b = b \log a$	$\log 8 = \log 2^3 = 3 \log 2$
	$\quad = 3 \times 0.301 = 0.903.$
$\log (a \times b) = \log a + \log b$	$\log 20 = \log (2 \times 10)$
	$\quad = \log 2 + \log 10$
	$\quad = 0.301 + 1 = 1.301.$
$\log (a/b) = \log a - \log b$	$\log 5 = \log (10/2) = \log 10 - \log 2$
	$\quad = 1 - 0.301 = 0.699.$
$\log (1/a) = -\log a$	$\log 0.0001 = \log (1/10{,}000)$
	$\quad = -\log 10{,}000 = -4.$
$\ln a = 2.303 \log a$	$\ln 100 = 2.303 \log 100$
	$\quad = 2.303 \times 2 = 4.606.$
$\ln e^a = a$	$\ln 20.085 = \ln e^3 = 3.$
$e^{(\ln a)} = a$	$e^{(\ln 20.085)} = e^3 = 20.085.$

The other laws for natural logarithms are similar to the laws for the common logarithms (base = 10) above. Converting a number to its logarithm entails the "gain" of one significant digit. For example, $\log (2.0 \times 10^5) = \underline{5.30}$, since the "5" comes from the "10^5" and the ".30" comes from the "2.0" (two sig. figs.).